The Gene Factory

The Gene Factory

INSIDE THE GENETIC AND BIOTECHNOLOGY BUSINESS REVOLUTION

JOHN ELKINGTON

CARROLL & GRAF PUBLISHERS, INC.
New York

First Carroll & Graf edition 1985

Carroll & Graf Publishers, Inc.
260 Fifth Avenue
New York, NY 10001

Library of Congress Cataloging in Publication Data

Elkington, John.
  The gene factory.

  Bibliography: p.
  Includes index.
  1. Biotechnology industries.   I. Title.
HD9999.B442E42   1985   338.4'76606   85-12812
ISBN 0-88184-208-7

Manufactured in the United States of America

Contents

Acknowledgements

Biotechnology is such a fast-moving field, and is proceeding on such a broad front, that it is probably impossible for any one person to keep abreast of all of it. I have visited many of the companies and organisations whose activities are described in *The Gene Factory* and owe a tremendous debt of gratitude to the literally hundreds of people who have given freely of their time and expertise. But I also owe a great deal to many others who, directly or indirectly, knowingly or unknowingly, have kept me in touch with developments around the world.

As Editor of *Biotechnology Bulletin* I have been supported by many people, but want to thank Nick Coles (as Managing Director of OYEZ Scientific and Technical Services Ltd), Deborah Hooper, Catherine O'Keeffe and, most particularly, Anita Urquhart. My editorial advisers have included Professor Eric Dunlop of the University of Washington at St Louis, Professor John Higgins of the Cranfield Institute of Technology and Dr Glyn Tonge of P.A. Technology.

A number of other publications have also helped support my developing biotechnology habit. Tim Radford of the *Guardian*'s Futures column will recognise more than most in the pages that follow. I have tremendously enjoyed writing for the *Guardian*, not least because of the response that some of the articles have drawn. I should also like to thank Christopher Edwards, the first Editor of *Bio/Technology*, and Terry Bishop of *Designing* for other commissions. The overwhelming bulk of the material presented here, however, has been accumulated while I produced successive issues of *Biotechnology Bulletin* and over sixty company reports published alongside it.

And there are other debts. The Department of the Environment's Central Directorate on Environmental Pollution commissioned my first work on biotechnology, in 1979–80. A Fellowship from the Winston Churchill Memorial Trust enabled me to undertake a long tour of the United States in 1981. Richard

Pearson of the Institute of Manpower Studies commissioned early work on the employment implications of biotechnology, during the Institute's pioneering study of the biotechnology brain drain for Britain's Science and Engineering Research Council. Dilmus James of the International Labour Office (ILO) commissioned a study of the employment implications of novel biotechnologies in the Malaysian oil palm industry (published in *Blending of New and Traditional Technologies*, Tycooly International Publishing Ltd, Dublin 1984).

OYEZ and the ILO independently funded two trips to Japan, in late 1983 and early 1984. The first tour resulted in a report published by OYEZ, *Bio-Japan: The Emerging Japanese Challenge in Biotechnology* (OYEZ, London 1984). And Gil Green of Online Conferences Ltd helped plug me into Biotech 83 and Biotech 84, the most informative biotechnology conferences I have yet attended.

The opportunity to write *The Gene Factory* came out of the blue, when Gail Rebuck, a Director of Century Publishing Ltd, asked me whether I would like to write a book on biotechnology. In retrospect, and having written books for a number of other publishers, I consider myself tremendously fortunate that it was Century. Gail helped knock the original concept into shape and Paul Barnett worked miracles in cutting the resulting typescript down to size.

The book would not have been written in the time available without the Burroughs B20 word-processing system which Gill Travis and Nigel Cope of Burroughs Machines helped me get operational. My major debt in this field, however, is to my colleague Jonathan Shopley, who fed the hardware with the appropriate software and chased bugs until he caught them. Thank you Jonathan.

Marek Mayer, with whom I worked for several years at Environmental Data Services Ltd, and who took over from me as Editor of the ENDS Report, helped shape my thinking on the environmental aspects of biotechnology. He has also fed me with various leads.

I have visited a number of companies and research organisations working in the biotechnology field as one of the Assessors for the innovative Pollution Abatement Technology Awards scheme, supported by the Confederation of British Industry, the Department of the Environment, the ERAS Foundation and the Royal Society of Arts. I would like to thank Timothy Cantell and Nancy Pace for their sterling efforts.

I have also learned a great deal from a number of writers who regularly cover biotechnology, most particularly from David

Fishlock, in the *Financial Times*, and Stephanie Yanchinski, whether writing in *Bio/Technology* or *New Scientist*.

But some of the most rewarding discussions I have had on the biotechnology prospect have been with my colleagues at Bioresources Ltd, including Phil Agland, Dr Conrad Gorinsky, Professor David Hall, Professor John Meadley, Nigel Tuersley and Dr Peter Waterman. These are indeed exciting times in biotechnology—and as one of the Directors of Bioresources I have experienced at first hand a few of the challenges facing those who seek to apply the new tools biotechnology is putting into our hands to some of the world's most pressing problems.

Finally, and most importantly, I should like to thank my family: my wife, Elaine, for reading the book as it emerged and for the many improvements she made in the text, and my daughters, Gaia and Hania, for all manner of support during the book's conception, gestation and delivery. By way of small recompense, *The Gene Factory* is dedicated to them.

Introduction: The Gene Factory

The whole building was shaking. Sitting next to me, the managing director of Kyowa Hakko Kogyo, one of Japan's leading biotechnology companies, turned and explained: 'It's an earthquake.' This was Tokyo in the late autumn of 1983, and the fact that the earthquake hit Kyowa Hakko's headquarters in the midst of a visit by American and European biotechnologists was singularly appropriate.

Neatly defined by Britain's Spinks Committee as 'the application of biological organisms, systems or processes to manufacturing industry and service industries', biotechnology has been the object of more public and commercial interest than any other emerging technology, with two exceptions: nuclear energy and the microchip. Its explosive growth, especially the mushrooming of the genetic engineering companies of North America and Europe, has shaken established companies like Kyowa Hakko to their very roots. For, if any country could have been said to have held a leading position in industrial fermentation in recent decades, it was Japan. In addition to its many traditional fermentation-based industries, such as brewing, distilling and soy-sauce manufacture, Japan had developed tremendous strengths in such fields as amino-acid and antibiotic fermentation. Yet suddenly, the Japanese found their lead coming under tremendous pressure from the new breed of biotechnologists, a number of whom were sitting around Kyowa Hakko's conference table that day.

Strikingly, during the first five months of 1983, a total of 172 presidents had stood down from top positions in Japanese companies, many of them because they felt they no longer understood the technologies on which the future of their companies would depend. 'I've read books on the introduction of biotechnology over and over, but haven't understood any of them really well,' was the way Tomu Tokosue plaintively explained his resignation as president of Teijin.

Japan's largest manufacturer of polyesters and its leading synthetic-fibre producer, Teijin had already signed contracts with two of the new genetic-engineering companies, Biogen (whose activities are described in considerable detail in Chapter 4) and Hybritech. The Biogen contract covered the development of new ways of producing commercial quantities of Factor VIII, a blood clotting agent used in the treatment of haemophiliacs (see pages 76–9), while the Hybritech contract focused on the use of monoclonal antibodies (see pages 28–9) in the treatment of certain types of cancer.

'Cancer,' said one Japanese magazine researcher. 'Everyone is scared stiff of it, including me.' Japan, in fact, leads the world as far as the number of deaths from stomach cancer is concerned, with some cancer experts suggesting that a major cause is the high salt content of many Japanese foods, including some which, like soy sauce, are produced by means of traditional biotechnologies. A few years ago, cancer overtook strokes as the country's No. 1 killer. The growing fear of the disease—which in Japan is rarely mentioned in public, even by doctors—has triggered massive growth in the markets for anti-cancer products and for cancer insurance. Not surprisingly, Japanese businessmen want a share of these new markets. But one of the recurrent nightmares from which Tomu Tokosue and other senior executives of Japanese companies have suffered centres on what they believe to be a very real possibility: that biotechnologists outside Japan will progress at such a rate that they will soon lock up most of the most valuable future markets with watertight patents.

'I paid a lot of money for that,' Kyowa Hakko's managing director, Dr Hirotoshi Samejima, said to me of his company's 1982 agreement with Genentech, America's leading genetic-engineering company, based in San Francisco. It covered research and development work on a product which looks like being a cure for heart attacks, one of the leading killers in the developed countries. Kyowa Hakko initially tried to develop this product, called tissue plasminogen activator (see page 100), on its own, but found that Genentech had already stitched up much of the field with a number of key patents. 'It was useless to compete,' Samejima explained, although his company has since been spending growing sums of money to ensure that it is not forever dependent on bought-in expertise in its pursuit of its key objective, 'formulating the future'.

And it is not just the older companies which are finding their businesses under threat from the emerging biotechnologies. Even a company like Amersham International, which enjoyed a spectacular, if controversial, launch on the British stock market as recently as 1982, has found its products—many of which are sold to the biotechnology industry itself—coming under growing challenge from alternatives generated by biotechnology (see pages 71–4).

During 1982, I visited many of the leading US biotechnology companies with Mike Brady of Amersham's radio-labelled compounds department. Not surprisingly, he took a very close look at the work under way in those companies—many of which recognised him as the 'competition' in some key emerging areas of their own businesses. I found those visits a dramatic revelation of the emerging challenge which biotechnology poses to existing companies and, indeed, existing industries.

Yet, surprisingly, over thirty years after the genetic code was cracked and more than a decade after genetic engineering was first shown to be feasible, the majority of people have not even heard of biotechnology—while few of those who have could give you a coherent idea of what it is or what its applications and implications might be. And this despite a flood of press articles and radio and television programmes, and the publication of enough books and reports to fill a sizable library.

So, perhaps not surprisingly, my reaction when first asked to write this book was: Not another book on biotechnology! And yet, while talking to people both inside and outside the biotechnology business, it struck me that most of the books currently available focus in considerable detail on the technologies and techniques used by biotechnologists; on specific issues, like the environmental risks and ethical issues associated with genetic engineering; or on particular applications, as in Brazil's emerging 'alcohol economy'.

The central objective of this book is to give the reader a sense of what it is like to work in the biotechnology business—and of the way in which biotechnologists see their processes and products shaping and reshaping our world.

As I write, there are plans afoot to make a movie about the cracking of the genetic code: indeed, James Watson, who discovered the double-helix structure of DNA (deoxyribonucleic acid) with Francis Crick, apparently suggested that Crick should be played by Roger Moore and himself by Dudley Moore! Whoever eventually plays the roles, the very existence of such plans is an indication that the story of that great discovery, first described by Watson in *The Double Helix* (1968), continues to capture the public imagination. The details of what Watson and Crick actually achieved, however, tend to be a great deal less distinct in the public mind.

In his book, Watson recalls that Crick raced into a pub, 'to tell everyone within hearing distance that we had found the secret of life'. What they had actually done, as a paper published in the April 25th, 1953, issue of *Nature* explained, was to reveal that the genetic information, the 'blueprint' of life found in all living organisms, is encoded in two long strings of just four chemical 'letters'—the nucleic acids which make up DNA—and that these strings are wrapped around each other to form a double helix. But Crick's characteristic outburst was more than justified.

In retrospect, that insight into the inner working of the living cell ranks alongside the discoveries of scientists such as Werner Heisenberg and Erwin Schrödinger, who opened up the heart of the atom. And, like early research on the structure of the atom, the new field of molecular biology has enormous potential for human welfare.

But, inevitably, such powers can be used for evil as well as good. It is worth recalling, for example, that while Schrödinger, not himself a Jew, made no secret of his loathing for the Nazis and was almost killed when he tried to interrupt storm troopers engaged in a pogrom, Heisenberg accepted high-ranking posts under the Nazis—and directed German research on the atomic bomb. Similarly, while the overwhelming majority of biotechnologists are earnestly pursuing research designed to exploit biotechnology's potential for increasing human welfare, others are investigating ways in which completely new weapons might be devised.

My own initial interest in genetic engineering was triggered by the growing public concern about the possible risks in terms of human health and ecological impact. Then, in 1980, I was asked by the UK Department of the Environment to investigate the environmental implications of a wide range of industrial technologies likely to feed through into the economy by the mid-1990s, biotechnology among them. It was only while carrying out that work—the results of which were later published as *Pollution 1990: The Environmental Implications of Britain's Changing Industrial Structure and Technologies*—that I began to appreciate the positive side of the equation.

During the 1970s, I had worked on research and consultancy projects in Europe and in such developing countries as Egypt, looking at the impact of development programmes on human health and on environmental quality. I had worked on projects focusing on the spread of parasitic diseases such as schistosomiasis; on the clean-up and treatment of some of the enormous quantities of toxic and often highly persistent chemicals which have been dumped into the environment; and on the exploitation of renewable energy sources. These, I now found, were just some of the many areas in which biotechnologists expected to make major and beneficial contributions.

I have become convinced that biotechnology's overall effect can be overwhelmingly positive, although there are of course many problems. Like all major technological revolutions, this one will result in major social and economic impacts. During 1983, for example, I carried out a study of the Malaysian palm-oil industry for the UN International Labour Office, concluding that the introduction of new oil-palm cloning methods and of biological pest control could generate unemployment. But the new tools afforded by biotechnology are already offering totally new employment opportunities by laying the foundations for new industries whose areas of business were unimaginable just a few short years ago.

Some of the world's brand-new gene factories are in trouble, however, because they are burning up their resources much faster than they can bring commercially attractive products to market. It is fairly easy to pull together the resources needed to launch a new start-up company, but very much more difficult, as Chapter 3 explains, to raise the finance to take any new products through the development, production and testing stages which must be negotiated before the first vial, packet or seed can be sold.

Today's investors are very much more choosy about the biotechnology ventures they will fund, yet the new companies will inevitably need growing amounts of money to scale up what are still in many cases little more than laboratory-bench technologies. It is easy to forget that a company like Ranks Hovis McDougall has had to spend £30 million to bring its fungal protein to the point where it is about ready to market (see page 152), while BP was forced to write off the bulk of its £100-million investment in single-cell protein technology (see page 175). ICI, which has still to make a profit on its Pruteen technology (see page 177), now sees its own £100-million investment simply as an 'entry fee' to commercial biotechnology.

Many of the new biotechnology companies will find market niches of a sort, but equally many will not. Indeed, there is growing apprehension in the industry that the honeymoon period is over. To begin with, the start-up companies were able to run rings around their established competitors—and were able to persuade many of those competitors to invest substantial sums for a 'window on the action'. But things are changing fast.

'The public has a distorted view of what's happening in biotech,' noted Jack W. Schuler, vice-president for biotechnology at Abbott Laboratories, one of the many major drug companies now moving strongly into this area. 'The small companies are extremely vocal in hyping their successes, and the large companies are saying nothing.' But the large companies are now intent on winning the next round. Some, like W. R. Grace, which plans to invest over $60 million in its agricultural joint venture with Cetus (see page 115), will choose to collaborate; others will simply buy up the companies they want, in the same way as Lubrizol has bought up Agrigenetics.

'It's becoming the waltz of the elephants,' as one financial analyst put it to *Business Week*, 'and the fleas are going to get squashed.' Ciba-Geigy, Du Pont and Monsanto are just some of the major companies which have opened massive new in-house biotechnology research facilities. 'The train,' the president of one start-up company admitted, 'is coming down the track.'

Whatever the problems that face us, however, these are truly exciting times, not least because of the advances now being reported by the world's biotechnologists. 'If you are young,' as James Watson put it in his opening remarks to a conference organised by *Nature* in Boston thirty years after the publication of that revolutionary paper, *A Structure for Deoxyribose Nucleic Acid*, 'there is really no option but to be a molecular biologist.'

While the overwhelming majority of us have no intention of becoming molecular biologists, it has never been more important that we should understand the keys to life's vital processes which molecular biology is now handing to us—and the locks which the biotechnology industry proposes to open with them.

ONE
The Genesis Experiments

Future archaeologists may stumble across the plans for a fusion reactor in strata carbon-dated to the late twentieth century and assume that we had abundant, cheap fusion power. Similarly, it may just conceivably be the case that a Babylonian tablet unearthed recently and dated circa 6000BC was an ancient's idea of science fiction, although one suspects it was anything but. Biotechnology has long roots.

The tablet depicts the stages in one of our most venerable biotechnologies, brewing. Several thousand years later, in the same corner of the world, the Sumerians, too, appear to have been much taken with biotechnology's intoxicating prospects. If their tablets are to be believed, they were brewing as many as nineteen different varieties of beer by the third millennium BC.

Our first, faltering steps in biotechnology must have come about as a result of a series of happy accidents—as in the discovery of champagne, much more recently, by the Benedictine monk Dom Perignon: his happy accident happened in 1668. Alexander Fleming's happened in 1928, as he scrutinised a culture of the fungus *Penicillium notatum*.

Accidents such as these are to be welcomed, of course, but early biotechnologists stretched every sinew in their attempts to understand the basis of life's diversity—and to manipulate it to their advantage. Despite their efforts, however, progress was painfully slow. Plant and animal breeders proceeded by the time-honoured process of trial-and-error. Even when a dramatic breakthrough occurred, it often took many years for its full import to strike home. In the case of Anton van Leeuwenhoek it took over a century. Peering through an elementary microscope in 1677, Leeuwenhoek discovered the single-celled organisms which we now know as protozoa. But he made perhaps his most extraordinary discovery in 1683, five years before Dom Perignon stumbled on champagne. Working at the limit of what his primitive, painstakingly hand-ground lenses could resolve, he described living structures which we now know to have been bacteria. It was to be more than a century before anyone else was to record seeing these minuscule micro-organisms.

Laying the Foundations

It is worth recalling, too, some of the other extraordinary developments which laid the foundations of modern biotechnology. Two men born in the same year, 1822, had a particularly profound influence. Louis Pasteur went on to become one of the most renowned scientists of all time while the Austrian botanist and monk Gregor Mendel died in 1884 without the slightest inkling of the fame which would be his after the Dutch botanist Hugo de Vries came across, in 1900, his long-forgotten paper on the laws of inheritance.

With astonishing patience, over a period of eight years from 1857, Mendel grew generation after generation of peas in his monastery garden. Hermetically wrapping his plants, to ensure that they were not randomly pollinated by passing insects, he painstakingly self-pollinated them. In this way he ensured that any characteristics the offspring inherited were from just the single parent. He hoarded the peas produced by each self-pollinated pea plant, later planting them and studying the effect of his experiments on each new generation. By careful cross-breeding trials, he uncovered the difference between dominant and recessive genes—without ever understanding exactly what a gene was.

Before his time, it had been widely believed that heredity was carried in the blood, and that the offspring's characteristics resulted from the blending of the blood of its parents. Mendel's experiments, by contrast, showed that some traits, such as the height of a plant, do not blend: crossing a tall plant with a dwarf plant does not automatically produce a medium-sized plant. Instead, he was able to show, the offspring's characteristics are inherited through factors which, far from blending, are actually segregated when the sex cells are formed. The extraordinary thing is that Mendel, knowing nothing of chromosomes or cell division, explained his results in a way which ties in very closely indeed with our current understanding of how hereditary mechanisms work.

Since then, the invention of improved microscopes and of new cell-staining techniques have made it possible to examine the cell nucleus itself, revealing the presence of what are now called 'chromosomes' ('coloured bodies'); this name derives from the fact that they stand out clearly when cellular components are stained. Closer study of chromosomes suggested that they must be the 'hereditary factors' predicted by Mendel—but later research showed that the chromosomes are themselves assemblages of such 'factors', now called *genes*.

Mendel wrote up his results for his local natural-history society, only to have them cold-shouldered. He also sent them to the Swiss botanist Karl von Nägeli, who delayed the development of the science of genetics for a full generation. Nägeli apparently looked at the paper but, as a biologist of the old school, was completely thrown by the mathematics. His dismissive comments discouraged Mendel, who immersed himself in the affairs of his monastery—of which he became abbot in 1868. To compound Mendel's problems, he became steadily fatter, which made it increasingly difficult for him to stoop down to the level of his experimental pea plants.

Pasteur's life seems to have been a great deal happier, not least because his many pioneering contributions to what we would now call biotechnology were recognised in his own time. He, like Mendel, suffered early poverty, as the son of a tanner, but he made his first major scientific breakthrough at the age of 26 and never really looked back. At the same time that Mendel was laying the foundations of genetics, Pasteur did much of the detective spadework which made industrial fermentation processes commercially viable.

In 1854 Pasteur became dean of the Faculty of Sciences at the University of Lille. There he soon became interested in the problems of the wine industry, one of the key supports of the French economy. Despite the industry's best efforts, wine and beer all too frequently emerged from the bottle sour and undrinkable, at enormous cost to the brewers and vintners. Many in the industry began to ask themselves whether the addition of some chemical might not solve the problem and, in 1856, one Lille industrialist approached Pasteur, by then a renowned chemist, in search of an answer.

At the time, many chemists were convinced that fermentation was a product of chemistry, not biology. But, by studying normal and soured fermentations through a microscope, Pasteur found that properly aged wines and beers contained spherical yeast cells, while soured vintages and brews proved to contain elongated yeast cells. Clearly, he concluded, there were two different types of yeast at work. One, it transpired, produced alcohol, while the other produced lactic acid. His solution to souring was the gentle heating of wine or beer to kill off any yeasts still left after fermentation. This process, now best known for its use in the milk industry, became known as 'pasteurization'. Although Pasteur's clients were initially appalled at the idea of heating their products, controlled trials demonstrated conclusively that the new process not only prevented souring but also left the quality of the wine or beer otherwise unaffected.

The Genetic Alphabet

The early 20th century is littered with biotechnology milestones: the first use of microbes for the large-scale treatment of sewage just prior to World War I; the development of bacterial ferment-ation for three critical industrial chemicals (acetone, butanol and glycerol) between 1912 and 1914; Alexander Fleming's discovery of penicillin in 1928; and the large-scale production of penicillin, in the build-up to D-Day, in 1944.

But the key event, the unlocking of the basic mystery of the gene, came with the discovery in 1953 of what Francis Crick unabashedly called the 'secret of life'. Although it would be twenty years before the announcement of the first successful genetic-engineering experiments, the knowledge that the struc-ture of the basic stuff of life, deoxyribonucleic acid (DNA), was a double helix was to be the keystone for the genetic-engineering industry.

In their short (900-word) paper announcing the 'double helix', Crick and Watson said little about the biological implications of their discovery, confining themselves to the observation that 'it has not escaped our notice that the specific pairing we have postulated immediately suggests a possible copying mechanism for the genetic material'. What did they mean?

Work with *Drosophila*, a fruit fly, had modified Mendel's theor-ies by showing that some genes appeared to be linked with one another, because they were located on the same chromosome, but that the strength of the linkage varied—depending on how close the genes were to one another on the chromosome. For example, white-yellow bodies, ruby eyes and forked bristles turned out to be linked traits, but the first two proved to stick together far more often than did either with the forked bristles. Research in the 1920s had also shown that genes are not perma-nent. While natural mutations generally occurred very slowly, they could be accelerated if fruit flies were exposed to X-rays.

Despite all this work, however, no one yet knew what genes were made of. The German physicist Max Delbruck had sug-gested that they might be chemical molecules, an idea which inspired such molecular biologists as Salvador Luria—whose own work did much to lay the groundwork for biotechnology. Then, in 1944, the same year that penicillin went into mass-production, Oswald Avery, Colin MacLeod and Maclyn McCarty of New York's Rockefeller Institute transferred DNA from one type of micro-organism to another, in the process proving beyond doubt that hereditary information is stored in the chemical structure of DNA.

Many different analogies have been developed to explain DNA and the new 'recombinant-DNA' technologies. Probably the most effective compares DNA to a library, containing the complete blueprint for an organism. Each of us results from a DNA library containing an estimated 3,000 volumes of 1,000 pages apiece. Each of these pages represents a single gene and contains about 1,000 letters. In contrast to the normal English alphabet, with its 26 letters, this chemical alphabet has only four letters (or 'nucleotide bases'): A (for adenine), C (cytosine), G (guanine) and T (thymine).

Most of us are now familiar with the basic model of DNA, a double-stranded, helical molecule, like a spiral staircase—with the 'rungs' consisting of pairs of these 'letters', A being always paired with T and C always with G. A gene, we now know, is spelled out in an ordered sequence of these chemical letters, with each gene containing the information needed to produce a particular protein, together with the chemical signals needed to initiate and halt its production.

The reason that Crick and Watson were so excited was that the double helix helped explain how DNA produces copies of such gene sequences. The pairing between the nucleotide bases is fairly weak, so that the process of cell division causes the DNA molecule to 'unzip' down the middle. The result is a pair of separate strands of DNA, each with a series of unpaired bases searching for replacement complementary bases. Since A will join up only with T, and C only with G, each strand thus serves as a template, and the end result is the formation of two identical DNA molecules.

But how are the instructions in this molecular code decoded? The decoding process, 'gene expression', involves two key steps: transcription and translation. In transcription, the DNA double helix is unzipped near the target gene and a single-stranded strip of messenger ribonucleic acid (mRNA) is synthesised. This mRNA strip is then released from the section of unzipped DNA which it has been copying and is used by the cell's protein factories, the so-called 'ribosomes', to produce the desired protein: this is the process called 'translation'. In the first stage of the exercise, then, the genetic message is copied, while in the second the resulting copy is translated into the language of proteins.

Each protein is composed of amino acids, of which there are 20 different types in living organisms, and each amino acid is coded for by three base pairs, called a 'triplet'. Once the protein has performed its allotted task, both it and the mRNA are degraded.

There are significant differences, however, between the genetic mechanisms of eukaryotes—the higher organisms (animals, plants and certain micro-organisms, primarily yeasts and moulds)—and the lower organisms, primarily bacteria, which are known as 'prokaryotes'. The essential difference between the two types of organism is that eukaryotic cells have a cell nucleus which contains a number of chromosomes, while prokaryotic cells have no cell nucleus and only one large chromosome, which floats free in the cell. Prokaryotic cells also contain small rings of DNA called 'plasmids'.

Bacteria, with perhaps 1,000 genes, are far simpler to work on than human cells, each of which may contain about a million genes. As we shall see later, however, there are some things which bacterial cells simply cannot do. A fair number of genetic-engineering companies, for example, are working on yeasts because they are closer to the cells of the higher eukaryotes. One key area of weakness in bacterial systems is that they cannot add vital sugar groups onto some of the most commercially attractive molecules.

The proteins produced by the sequence of transcription and translation processes are responsible for most of a cell's basic functions. The most diverse group of proteins is made up of enzymes, which are in effect biological catalysts, initiating and speeding biological reactions. A second group, the structural proteins, is used to build cell membranes and other elements of the cell. A third group, including the hormones, helps regulate cell functions, while others, such as haemoglobin, have other specialised tasks—in the case of haemoglobin, the transport of oxygen from the lungs to the rest of the body.

Luckily for today's genetic engineer, the genetic code is universal: all organisms use it. Since one of the key activities in genetic engineering involves coaxing DNA from one organism to express in a totally different organism, this fact is clearly critical. There are, however, important differences in the way the DNA molecules of different species code the 'start' and 'stop' regulatory signals which control gene expression. One of the most important activities in genetic engineering, then, involves hunting for the appropriate signals to stick onto the beginning and end of a sequence of DNA which is being inserted into a foreign host organism.

In the early 1960s, scientists uncovered ways in which *natural* genetic transfer can occur between bacteria, work which led directly to many of today's most powerful genetic-engineering techniques. One such route involves bacteriophages, which are viruses that infect bacteria. Like so many hypodermic needles, phages inject their DNA into bacterial hosts, which then pass it on to future bacterial generations as part of their own DNA. Occasionally, however, this viral DNA suddenly enters an active phase and produces a new crop of viral particles. Some of these may burst out of their host, often carrying particles of the bacterium's own DNA along with them. So when these fugitive particles infect another bacterium they may bring along genes from their previous host.

Another route involves the direct transfer of genes in a process called 'conjugation', whereby one bacterium attaches a small projection to the surface of another bacterium, and the DNA passes through the projection from the donor to the recipient. This ability is genetically controlled, but the relevant genes are found not on the bacterium's chromosomes but on those independent genetic elements, the plasmids: these are so small that they can pass into and out of cells with relative ease.

Not surprisingly, then, phages and plasmids were eyed with growing interest by those who wanted to transfer *particular* bacterial genes as and when required.

Genetic Scissors and Paste

But there was a fly in the ointment: bacteria proved to have evolved a means of dealing with rogue DNA, however it found its way into them. Bacteria have evolved a range of 'restriction enzymes' which cut DNA molecules where specific sequences of nucleotides are found, enabling them to slice up intruding DNA while leaving their own intact. An important discovery came when Salvador Luria had another of those extraordinary accidents which have led to major advances in the biosciences. He discovered the enzymes that splice DNA when he broke a test tube containing one bacterium and borrowed another which contained a culture of a different strain. But then, as Pasteur once put it, 'chance favours the prepared mind'. Luria's work was developed by such leading scientists as Herbert Boyer (who was later instrumental in setting up Genentech), Stanley Cohen, and Paul Berg (who helped trigger the debate about the risks involved in this type of work).

Today restriction enzymes, or endonucleases, are the 'scissors' used by genetic engineers to open up plasmids and insert foreign DNA. Another group of enzymes, the ligases, provide the 'paste' used to stick everything back together again. If a nucleotide sequence occurs only once on a plasmid, then the appropriate restriction enzyme will open up the plasmid just once, like a child cutting through a hoop with a pair of scissors, but, if the sequence occurs a number of times around the plasmid, then the DNA will be chopped up into several pieces.

By the late 1970s, scores of restriction enzymes had been isolated from bacteria, some being found in the most unlikely places. New companies, like P & S Biochemicals, were set up to hunt for these enzymes and to market them as tools of the biotechnology trade. As a result, the cloning of DNA—which involves producing a large quantity of a given DNA molecule by inserting it into a host bacterium—became increasingly routine. The first successful cloning experiment took place in 1973, while the following year saw the first expression of a gene cloned from a different species of bacteria.

This is not the end of the story, however. Once inside the host cells, the recombinant-DNA (rDNA) plasmids replicate themselves over and over again. In the process, they also replicate the lengths of foreign DNA which have been spliced into them. But, because the restriction enzymes have typically cut out various different DNA sequences, only a few of the plasmids will be reproducing the desired DNA fragment.

So how can we track down the target sequences? One widely used method for screening recombinants involves the use of the industry's 'workhorse', a plasmid called 'pBR 322'. This is only 4,300 bases long and contains genes which make the host cell resistant to two antibiotics—tetracycline and ampicillin. If foreign DNA enters either of these genes, then the specific resistance it confers is lost. Theoretically, the resulting clones could be of three distinct types. One would be resistant to both antibiotics, showing that the plasmid had been inserted unchanged. A second group would be resistant to just one antibiotic, indicating that a plasmid carrying foreign DNA had been inserted. And the third group would show no resistance to either antibiotic, showing that the plasmid had not been accepted by the host cell. This system, where appropriate, can speedily identify clones which have accepted the foreign DNA.

Another answer is called a 'DNA probe', a sequence of DNA which is identical to the target gene sequences but has been labelled in some way, so that it can be identified once it has paired with them. Such DNA probes can now be produced to order with DNA synthesizers, machines that can be programmed to assemble nucleotide bases into tailor-made DNA sequences. (A more detailed description of this type of equipment, the 'gene machine', can be found on page 57.)

Coming up with an effective DNA probe can sometimes be the hardest part of the cloning process. The first genes to be cloned typically produced large quantities of mRNA, which meant that the mRNA could be used as a DNA probe. Most genes, however, do not produce such quantities of mRNA, and so other approaches have had to be developed. One possibility involves determining the amino-acid sequence of a protein and, working backwards, finding the sequence of the gene's nucleotide bases—and then using a DNA synthesizer to produce a complementary stretch of DNA.

Some complete genes have been synthesized, including that for the hormone somastatin, a tiny protein just 14 amino acids long; with three nucleotides coding for each of those amino acids, a DNA chain 42 nucleotides long had to be synthesized. This approach is still, not surprisingly, out of the question for more complicated proteins. By 1984, gene synthesis took about fifteen minutes per base and the largest DNA sequences produced routinely were about 60 bases long. The pace of development should not be underestimated, however. In 1983, for example, Andrew Murray of the Dana-Faber Cancer Institute and Jack Szostak of Harvard Medical School managed to synthesize the world's first working artificial yeast chromosome. Initially, however, this artificial chromosome proved to be less stable than its natural counterparts, which were ten times longer.

Once you have produced a length of DNA, by whatever means, you still have to determine that it is the right one, through a process known as 'sequencing'. To do this, the usual approach is to mark one end of the molecule with a tag of radioactivity and then to subject it to a series of reactions which interrupt it in a predictable fashion. Radioactive DNA fragments of varying lengths are thereby produced, and these can be separated through a process called 'gel electrophoresis'. This sequencing process produces a developed X-ray film. If you know how the reactions break up the molecule and, by reading the developed X-ray film, the size of the DNA fragments you have generated, then you can determine whether your product has the right sequence of bases.

In its 1982 annual report, the US company Genex summarised the key elements of the genetic-engineering business. It explained:

DNA can be thought of as a language, the language in which all of nature's genetic information is written. As with any language, it is desirable to be able to read, write and edit the language of DNA. Rapid methods for determining the substructure of DNA (DNA sequencing), developed a half a dozen years ago, correspond to *reading* DNA. These methods now make it possible to determine the complete structure of a gene in a few weeks. New and still rapidly evolving methodologies for chemical synthesis of DNA molecules make it possible to *write* in the language of DNA much more rapidly than was possible only a couple of years ago. Finally, and most important, genetic-engineering techniques themselves (recombinant DNA methodology) make it possible to *edit* the language of DNA. It is by this editing process that the naturally occurring text can be rearranged for the benefit of the experimenter.

Genex also stressed that 'new words (genes) can be introduced into the text, and new and more suitable texts (organisms) can be chosen. In addition, the editing process allows more subtle changes in the DNA molecule to be made, changes that can result in focusing much of the cell's metabolic energy on producing a single, specific protein from a single gene of choice.'

As far as the 'texts' go, biotechnologists have been working on a number of alternatives to their ubiquitous workhorse, the bacterium *Escherichia coli*. Given the fact that *E. coli*'s normal habitat is the human intestine, it is hardly surprising that its widespread use in experiments designed, say, to transfer antibiotic resistance has triggered concern. The alternative host organisms now being worked on include *Bacillus subtilis*, which offers the genetic engineer a number of important advantages over *E. coli*. For a start, it lacks the endotoxins found in the *E. coli* cell wall; these endotoxins, which are integral components of the cell wall and are released only when the wall is disrupted, can cause fevers and have helped slow the regulatory approval of products extracted from *E. coli*. The use of *B. subtilis* can dispense with the need to carry out complicated and time-consuming tests for such endotoxins.

There are other advantages. Extracting products from cells once fermentation is complete is generally a fairly fraught business (as Chapter 10 explains). When the products you want are produced inside the bacterial cell and trapped there, you face a number of problems. Because of their small size and relatively tough cell walls, bacteria are highly resistant to most efforts to fragment them. The more vigorously you try to mash up the cells, the more you risk damaging (or 'denaturing') the very product you are trying to extract. Gentler alternatives are either inefficient and time-consuming or expose the protein to attack by other bacterial enzymes.

In contrast to *E. coli*, however, which typically excretes just 20 per cent of the target proteins, *B. subtilis* has long been known for its production of proteins outside its cellular surface. It also excretes such proteins into the fermentation broth, which can make them easier to separate. New strains have been developed which are even more efficient, although supporters of *E. coli* are unlikely to give up without a fight. A Japanese team at the Institute of Physical and Chemical Research (RIKEN), for example, has produced a strain of *E. coli* which excretes 90 per cent of the target proteins.

There are even those who believe that many genetic engineers may one day put all their bacteria, yeasts and mammalian cells back on the shelf and switch to using insect cells as hosts for their experiments. Max Summers and Gale Smith at Texas A&M University hooked the gene for human beta interferon onto a virus which, in turn, was used to infect cells from the ovaries of army fallworms. These cells not only gave high yields of interferon, but produced it in the glycosylated form (that is, with the sugar groups which bacteria cannot attach to such molecules) and secreted the product into the culture medium—facilitating separation and purification.

A Nobel Factory

While the biotechnology companies have been developing and exploiting this wealth of new tools, basic research institutes around the world have continued to unravel more of life's deepest mysteries. To get a better idea of some of the trends, I visited the Medical Research Council's Laboratory of Molecular Biology, one of the most prolific of the world's bioresearch facilities and the organisation in which the genetic code was finally cracked.

Indeed, if molecular biology is the foundation stone of commercial biotechnology, then the Laboratory of Molecular Biology (LMB) could claim to be the foundation stone of molecular biology. Now located at the University Medical School on the outskirts of Cambridge, the LMB has hosted a succession of Nobel laureates: Francis Crick, John Kendrew, Max Perutz, Frederick Sanger, James Watson, Aaron Klugg, Cesar Milstein and George Koehler. Today it remains at the forefront of molecular biology, pioneering in such fields as monoclonal antibodies (invented by Cesar Milstein), molecular graphics (pages 216–19) and 'protein engineering'.

Like biotechnology itself, the LMB has long roots. The Medical Research Council (MRC) set up a unit for research on the molecular structure of biological systems in 1947, essentially as a vehicle for Max Perutz and John (later Sir John) Kendrew to continue their work on the use of X-ray diffraction techniques for the study of protein structures. At the time, the use of this technique to uncover the three-dimensional architecture of such large molecules as proteins and amino acids seemed, in the MRC's own words, 'a hopeless undertaking'. Such molecules are orders of magnitude larger and more complicated than the sugars and vitamins whose atomic structures were then known.

The double-helix work of 1953 sent a surge of adrenalin through the LMB—and ensured that Crick and Watson shared the 1962 Nobel Prize for Physiology or Medicine with Maurice Wilkins. 1953 also saw a key breakthrough in X-ray diffraction, which involves the interpretation of interference effects caused when the rays pass through a crystalline structure. Max Perutz developed a way to decipher diffraction patterns from crystalline proteins. The heavier an atom, the more efficiently it diffracts X-rays, and Perutz therefore added a single atom of a heavy metal like gold or mercury to each protein molecule, greatly simplifying the process of solving protein structures.

This new method allowed Kendrew and his collaborators to solve the structure of myoglobin, a protein which stores oxygen in muscle, while Perutz himself solved the structure of haemoglobin, the oxygen-carrying protein found in red blood cells. For this work, Perutz and Kendrew shared the 1962 Nobel Prize for Chemistry. X-ray diffraction has since led to the unravelling of the structures of over 100 different proteins, although very few of these are likely ever to have much commercial significance.

Sydney Brenner, now LMB director, joined the Unit in 1956 and, with Crick, carried out work on bacteriophages and bacteria which ultimately led to the uncovering of the triplet nature of the genetic code and to the discovery of messenger RNA.

All this work was done while the LMB was still housed at the Cavendish Laboratory. Meanwhile, in Cambridge University's Department of Biochemistry, Frederick Sanger was trying to solve another seemingly insoluble problem: the chemical formulae of protein molecules. Sanger was basically interested in determining the exact structure of the amino-acid chains of such molecules and, after eight years of intensive work, he worked out the precise sequence of amino acids in the entire insulin molecule. He was awarded the 1958 Nobel Prize for Chemistry.

As has so often been the case with LMB research, this work was a tremendous stimulus for other groups: Choh Li's team at the University of California solved the structure of the pituitary hormone ACTH, for example, while Vincent du Vigneaud not only solved the structure of simpler molecules like oxytocin and vasopressin but also moved on to synthesize them. Oxytocin was the first hormone ever synthesized, and so du Vigneaud was awarded the 1955 Nobel Prize for Chemistry. Li went on to isolate human growth hormone in 1956 and, in 1970, actually synthesized it, despite the fact that it was made up of 256 amino acids.

The commercialisation of such products as recombinant human insulin and human growth hormone was still years in the future, but these new methods of determining the structure of proteins and nucleic acids, coupled with advances in microbial genetics and biochemistry, had effectively opened up a complete new field of research, molecular biology. Realising its potential, the MRC decided to build a new laboratory to merge Perutz's and Sanger's work. The result was the LMB, which opened in 1962.

Sanger's new methods for determining the sequence of bases along nucleic acids allowed genes to be deciphered, enabling scientists to read the language of DNA. At the time, little was known about the chemical structure of genes, beyond the nature of the genetic code itself. Sanger's work opened up a new world of genetic grammar and syntax, changing the face of genetics. Before his research on nucleotide sequencing, geneticists had had to infer the structure of genetic material from genetic crosses, whereas they can now use direct chemical analysis. His methods have been applied worldwide, so that nearly two million nucleotide sequences are now available, including the complete sequences of some viruses.

Much of the LMB's research still centres on the determination of the structures of large biological molecules and on the study of their functions. But its scientists are now able to analyse much more complex biological systems than those represented by simple viruses and bacteria. The LMB has expanded into cell biology, immunology, developmental biology and neurobiology. Without question, such institutes are producing the seed corn of tomorrow's biotechnology, but their financing, given the essentially long-term nature of most of their research, can be a problem. Industry is typically not very interested in investing in basic research of this type (although Chapter 2 looks at some of the exceptions). Ironically, as LMB director Sydney Brenner had been at pains to point out, the value of the Laboratory's budget has been shrinking in real terms, with its capital expenditure having been cut by more than 50 per cent. And this at a time when molecular biology is acknowledged as a key ingredient of successful biotechnology.

A Flood Tide of Antibodies

One area in which there has been considerable industrial interest in the LMB's research has been the development of monoclonal antibodies. These have caused almost as much excitement as the basic genetic-engineering techniques already described. Immunologists study the immune response (see page 90), which protects all higher organisms against viruses, bacteria and other health challenges. A key element in our immune response is the production of antibodies by specialised cells called 'B-lymphocytes', found in the spleen, lymph nodes and blood. These recognise substances which are foreign to the body (antigens), and produce antibodies which bind to them. When a particular B-lymphocyte encounters and recognises an antigen for the first time, the B-lymphocyte is programmed to produce the appropriate antibody for the rest of its life.

This fact is the basis of all vaccination programmes, which challenge our immune systems with a small dose of a dead or weakened version of the virus or bacterium. Conventionally, antibodies have been produced by injecting an antigen into a laboratory animal. As a result, the animal's immune system produces a surge of antibodies, which are then harvested by collecting antiserum (i.e., blood serum containing the relevant antibodies) from the animal. Among the problems which have hampered this approach have been the contamination of the injected antigen, resulting in a cocktail of antibodies being harvested, and the fact that different animals do not necessarily produce totally homogenous antibodies when exposed to the same challenge.

The breakthrough came when Cesar Milstein and George Koehler began experimenting with cell-fusion techniques, fusing antibody-producing tumour (myeloma) cells with antibody-producing B-lymphocytes which had been immunised with red blood cells extracted from sheep. They found that the resulting hybrid cells (hybridomas) produced large quantities of identical (monoclonal) antibodies against the sheep red blood cells. The myeloma parent cell, extracted from a tumour, had conferred its own immortality on the hybridoma, permitting it to grow indefinitely in cell culture, producing almost unlimited quantities of antibody. The B-lymphocyte parent cells, meanwhile, had contributed the genes coding for the specific antibody.

However, just as the clones resulting from a gene-splicing exercise have to be screened for a desired DNA fragment, so hybridomas had to be screened for a desired monoclonal antibody. Then, in 1984, Nova Pharmaceutical, based in Morristown, New Jersey, announced that it had licensed exclusive worldwide rights to a patent-pending breakthrough which promised to increase massively the productivity of monoclonal antibody production. The new technology, developed at Johns Hopkins University, promises to increase yields of hybridomas from the typically low range of between one and five per cent to virtually 100 per cent. So, instead of having to screen their hybridomas, researchers will know that the overwhelming majority are producing the desired monoclonal antibody.

The first company to clone and express a complete antibody molecule in a bacterial system was Genentech, in 1983, working with scientists at the City of Hope National Medical Center in Los Angeles. The new monoclonals, Genentech announced, should be cheaper and should avoid 'serum sickness', sometimes caused by an immune reaction to foreign proteins contained in products extracted from horse or mouse serum.

At the LMB, Milstein's group has for many years studied the molecular basis of the immune response, trying to find out how an organism can produce a seemingly unlimited variety of specific antibodies in response to specific antigens. Initially they determined the overall chemical structure of the antibodies, but increasingly the emphasis has been on the structure of the genes responsible for antibody synthesis, and on the changes they undergo during the development of antibody diversity. One outcome of the group's work was the invention of a method of producing pure monoclonal antibodies; this method was used by David Secher to develop the first monoclonal antibody to interferon, one of the most popular early products pursued by the new biotechnology companies.

The antibody was used to purify interferon preparations so that their biological and clinical properties could be investigated. Trials at the MRC's Common Cold Research Unit showed, for the first time, that pure interferon can protect people against virus infections. Later the antibody was used to develop new methods for assaying interferon levels. This 'immuno-radiometric-assay' technology was licensed to Britain's 'flagship' biotechnology company, Celltech, whose anti-interferon products had achieved sales of about £1 million by mid-1984.

But the research which has attracted the most recent media attention has been Dr Greg Winter's work on 'protein engineering'. This project, undertaken in Milstein's division of the LMB, involved altering the fine structure of a naturally occurring enzyme, tyrosyl tRNA synthetase. The project (described in Chapter 12) boosted the performance of the enzyme, specifically improving its ability to combine with its substrate, adenosine triphosphate (ATP). The long-term commercial implications of this technique could be very significant indeed, and companies like Genex have been investing a great deal of effort in protein-engineering programmes. Winter warned, however, that his team might have been lucky with the particular enzyme they chose to work on—and that it might cost several million pounds to improve an industrial enzyme using this technique.

Many new techniques are now streaming out of institutes like the LMB and out of the R&D units of many of the new biotechnology companies. (An area which we have not even touched on yet, for example, is the development of genetic-engineering techniques for use in plants; see Chapter 6.) Overall, the technology is becoming more powerful by the day. But, every so often, one comes across someone who displays surprising modesty about what has been achieved to date.

Genetic engineering, Sydney Brenner told delegates at a conference celebrating the 30th anniversary of the discovery of the double helix, 'is more than we are doing now—engineering contains the essence of design'. Today's molecular biologists and genetic engineers can manipulate genes happily enough, but still lack a complete understanding of the array of molecular and cellular mechanisms involved even in basic cell functions. 'Perhaps,' Brenner mused, 'we are not genetic engineers today, but only genetic mechanics.'

TWO
Living Inventions

One day, like latter-day Rembrandts and Picassos, genetic engineers may actually *sign* their work with a flourish of 'nonsense DNA'. This is a term used to describe the large quantities of DNA which, despite being found in genes, do not appear to have any immediate influence on the manufacture of proteins. Ultimately, of course, we may find that this material codes for the folding of proteins or some other key activity, but genetic engineers will always be able to lay their hands on DNA which is mute when inserted into an artificial gene. Genetic plagiarists, unable to decide what is essential in an engineered gene, will tend to copy the lot, the 'signature' included. Enter the law.

Some parts of the biotechnology industry have recently been devoting almost as much energy to protecting their existing genetic creations as to developing new ones. The new companies are now making as much noise about hiring top-flight patent lawyers as they once did about recruiting cloners and splicers. Copyright lawyers have engaged in tortuous arguments about whether novel gene sequences can be copyrighted. The US Constitution, for example, authorises Congress to 'promote the Progress of Science' by 'securing for limited times to Authors' the exclusive right to their "Writings"'. The Constitution, bent this way and that by the legal profession, has proved astonishingly flexible. Previous rulings in the US courts have shown that, while you may not be able to copyright a tape cassette, a floppy disc or a gramophone record, you can copyright the information recorded thereon, whether it be computer software or rock music. But, biotechnologists have asked, what about the genetic information 'recorded' in a plasmid?

One answer runs like this: When you splice a new gene into a plasmid, together with your best 'start' and 'stop' signals, you have produced an 'original work of authorship'. So, some lawyers advise, there is nothing to stop you copyrighting your favourite stretch of DNA. Some brave company may yet try its luck in the courts, but in the meantime biotechnologists are relying on patents and patent applications to protect their commercial flanks.

The Chakrabarty Furore

When the US Supreme Court ruled in 1980 that a man-made micro-organism could be patented, the name 'Chakrabarty' hit the headlines—and the decision was variously described as having 'assured this country's technology future' and as heralding 'the Brave New World that Aldous Huxley warned of'. The idea behind a patent is that it grants legally enforceable property rights in an invention to the invention's owner. Patents can now be awarded on new biotechnology *products* (including new micro-organisms and plasmids), new *processes*, new *product composition* and new *methods of use*. However, the idea that genetic engineers were not only going to be out there creating lifeforms, described during the Chakrabarty case as a 'gruesome parade of horribles', but also patenting them proved too much for many people. There was uproar.

Yet, as Dr Thomas Murray has commented

> we routinely allow patents on certain botanical varieties— organisms much more complex, and perhaps even more important economically than Chakrabarty's. We also allow people to own breeding stocks of mammals—cows, horses, sheep and other animals important to humans. If we see no problem with someone owning a prize bull, and in that way exercising control over that animal's activities, especially its reproductive future, how can we object to the molecular biologist's owning another pool of commercially desirable genetic material?

Ananda Chakrabarty was a General Electric research scientist when, in 1972, he developed a novel strain of bacteria that could break down four of the main components of crude oil, which pollutes beaches around the world. The way he did this was to extract, from a number of different *Pseudomonas* bacteria, plasmids which each gave the original parent cell the ability to break down a single component of crude oil. He then inserted these plasmids into a single strain of bacteria, in the hope that the resulting super-bug would degrade oil spills into harmless by-products and then disappear.

The super-bug worked well enough, although it has never been used in the real world. But Chakrabarty, meanwhile, had a problem. A key difference between a work of art and a genetically engineered micro-organism is that the latter can reproduce itself —generally at a fearsome rate. Anyone who has developed a recombinant microbe able to produce insulin or interferon knows that it is a long, high-risk and extraordinarily expensive exercise —yet, once the thing exists, anyone who manages to track down a culture of it can blithely begin harvesting its product without having had to pay all the original overheads that have driven *your* price up.

Many companies, because they sell only the end-product, are confident that they will be able to keep their microbes under wraps, as closely held trade secrets, and are therefore not bothering to apply for patent cover. Chakrabarty's microbe, in contrast, was tailor-made for release into the environment, where anyone could scoop up a handful of the active ingredient and race off to set up a competing clean-up operation. If Chakrabarty wanted to make money from his labours, he had no option but to apply for a patent.

So he applied to the US Patent Office and was duly awarded a patent on the process by which the bacterium had been developed and on the delivery mechanism, which involved mixing the active ingredient with straw. But the Patent Office refused to grant a patent on the bacterium itself, ruling that the only living organisms which were patentable under existing law were plants (bacteria are generally classified as neither plants nor animals, but as Protista). Chakrabarty lodged an appeal and, in a landmark decision, the Patent Office concluded that the inventor of a genetically engineered micro-organism, whose invention otherwise met all the legal requirements for obtaining a patent, could not be denied one just because his invention happened to be alive.

The ruling led some critics to argue that patents stultify innovation, allowing large corporations to 'lock up' any inventions which could threaten their established businesses. Ironically, however, biotechnology itself provided a key case refuting much of this argument. We have seen that it took 16 years to bring that 1928 invention, penicillin, into large-scale production, despite its huge promise as a weapon against bacterial infections. Alexander Fleming fought for ten years to get the money and facilities to produce and purify penicillin, but it took World War II to really get things moving. Sir Howard Florey, who shared the Nobel Prize with Fleming for the development of penicillin, later remarked that the delay in getting this medical miracle-product into large-scale production resulted from their failure to patent it. He called this failure 'a cardinal error'.

Historical precedent or no, the biotechnology industry has clearly decided that patent protection is worth having. Companies have been churning out thousands of patent applications. Genentech filed more than 1,400 patent applications up to early 1984—and by that stage had received a grand total of 80 patents. Biogen, Genentech's arch-competitor, was saying that it would spend $1 million that year on patenting fees. In the USA fewer than 100 biotechnology patents had actually been issued, although well over 1,000 applications were pending.

The Stanford Cliff-Hanger

But the patent applications which really held the industry in thrall were ones submitted by Stanford University's Technology Licensing Office. 'We don't know why they're withdrawing the issuance of the patent,' said the office's director, Neil Reimers, shortly after the news broke that the Patent Office was not going to issue a Stanford bio-patent on the scheduled date of July 13th, 1982. Certainly the Patent Office's reasons for slamming on the brakes had not then been made public, but Mr Reimers was less than accurate when he went on to say that the news was 'not of great concern'. It was, not only for Stanford but for the entire industry.

The patent in question was the second applied for by Stanford and the University of California in respect of the pioneering genetic-engineering work carried out by Stanford's Stanley Cohen and by Herbert Boyer of the University of California, San Francisco. The first patent was granted in December, 1980, and a total of 73 companies had each paid a $10,000 licensing fee by the end of 1982. By that stage, the two universities had earned about $1.5m in total, small beer in relation to the billions of dollars some expected they would earn if both patents came into full force.

The Patent Office's rejection of the second Stanford patent application was a matter of considerable concern not only because of the possible loss of revenues but also because the first patent, in principle at least, was vulnerable to challenge on the same grounds (see below), given that it was based on the same work. Indeed, some companies with a long-standing interest in biotechnology had never signed up for licences, in the belief that the patents would ultimately prove to be unenforceable. A key non-signer was the oil company Exxon—and it was an Exxon lawyer, Albert Halluin, who initially rocked the boat by pointing out some of the inadequacies in the second patent application.

Even so, until the second application was rejected it had looked all but inevitable that Stanford would dominate genetic engineering. Its first patent was considered the broadest in the biotechnology field and Stanford had frequently stated its belief that it would prove as significant to genetic engineering as Bell Laboratories' patent on the transistor had been in the development of the semiconductor industry. The work was reported in a paper, published in 1973, written by Drs Boyer, Annie Chang (of Stanford), Cohen and Robert Helling (of the University of Michigan). This described the successful construction and replication of a recombinant plasmid capable of transferring antibiotic resistance into E. coli.

For the first time, this research had demonstrated the feasibility of gene transfer. However, because of the prior publication of some of the research results, this first patent suffered a number of limitations. Although the US Patent Office issued a patent, other national patent offices did not. As a result, there had been little to prevent a foreign company using the patented process overseas and then importing the resulting products into the US market.

The second patent application, the 'product' application, was designed to reinforce the basic 'process' patent by patenting the genetically engineered plasmids used to transfer genetic material from one cell to another in the recombinant-DNA process. If Stanford relied on the first patent, it knew it would have to visit the laboratories of unlicensed companies to check whether they were using the patented method. Clearly, a 'product' patent would be very much easier to enforce.

Stanford had gone to considerable lengths to ensure that the terms of the licensing agreement were acceptable to a large proportion of commercial genetic engineers. The annual fees paid prior to the launch of any product, for example, could be counted against the royalties for that product, provided the licensee signed up before December 15th, 1982.

But then the legal spanner was thrown into the works. While the Patent Office came up with a variety of technical reasons for rejecting the application, two key problems centred on the issue of premature publication and 'sufficiency'. Under patent law, inventors in the USA have one year after the publication of their findings to apply for patent protection. Sadly for Stanford, an article was published in New Scientist on October 25th, 1973, one year and one week before the patent application was filed. The 'sufficiency' argument looked more serious, however, revolving around the question of whether the patent described the relevant plasmid sufficiently to permit other scientists to reproduce the invention.

Stanford had to prove that the article did not provide enough information for reproducing the invention but that the patent did. A central problem as far as the latter part of the required proof was concerned was that, while the first Stanford patent described a process for producing the plasmid pSC101 from the known plasmid R6-5, Dr Cohen later admitted that pSC101's parentage was untraceable—and that he had failed to inform the Patent Office of the fact. Stanford played down the importance of the plasmid, saying that it had been made available to *bona fide* researchers, so that its parentage was not critical.

Stanford's response to the Patent Office ended by stressing that 'this is an uncommon invention. Few inventions spawn industries of such magnitude so quickly. Few inventions provide such sophisticated, sensitive and productive tools to investigate an incredible array of physiological processes and to permit the production of an infinite variety of products'. Few genetic engineers would have disputed this contention, but a prolonged period of negotiation ensued, with Stanford answering a succession of challenges from the Patent Office.

Rumours continued to surface every few months during 1983 that the second patent would be granted at any moment, but 1984 came and it was still in the pipeline. By this time Stanford had taken in $2.7 million. Later in the year, it sent a letter to its licensees, informing them that the patent was on its way, but would cover a very much narrower field. More specifically, it was expected that it would cover only products produced by bacterial plasmids in bacterial hosts. Finally, after ten years of controversy, the product patent was awarded on August 28th, 1984, with some analysts predicting that the two universities could earn over $1 billion. The patent expires in 1997.

Despite Stanford's problems, however, other universities were soon vying to get their own patented technologies into the market-place. In the USA, Harvard and the Massachusetts Institute of Technology (MIT) have been racing to move out of Stanford's shadow. One of Harvard's key patents covered a process invented by Nobel Laureate Walter Gilbert, who resigned his tenured position at Harvard to join Biogen. We saw on page 24 that the tough walls of bacterial cells can be a problem when the time comes to extract the protein they have been induced to make. Gilbert's team worked out a way around this problem. This involved hooking up the target protein with another protein which, in normal circumstances, is excreted from the bacterial cell. If everything goes to plan, the protein which you actually want emerges from the cell 'piggy-backing' on the excreted protein. This Harvard patent covered *E. coli*, *B. subtilis* and yeasts.

But Harvard and many other universities found it increasingly difficult to sell licences for such technologies. Partly this was because the rapid rate of innovation meant that it was only a matter of time before someone else invented their way around the patented process, but there were other reasons, too. One was the scepticism generated by the Cohen-Boyer licensing programme operated by Stanford. When those licences were first announced, explained Harvey Price, executive director of the Industrial Biotechnology Association, 'everybody figured this was it. If they wanted to be a player in the gene-splicing game, they had better take a license.' But as more and more patents were issued, many companies which had signed up for the Cohen-Boyer licences realised that, if universities and competing companies kept taking slices out of their long-term revenue stream, there would be little left to justify the whole exercise. Something had to give.

One proposal suggested to get around this problem was 'patent pooling'—an idea devised by Neil Reimers of Stanford and Roger Ditzel, director of the University of California's Patent, Trademark and Copyright Office—devised, in short, by those who had most to gain from the Cohen-Boyer licences, if they held. The basic concept was that anyone wanting to get involved in genetic engineering could pay a single licensing fee to the patent pool, covering all the necessary tools of the trade, with the inventors of those tools taking their return out of the pool, rather than insisting on separate negotiations with each user.

The problem, at least as far as inventors were concerned, was that, while industry would almost certainly make more use of the relevant technologies under this proposed new arrangement, those who had developed them stood to make less money. Those likely to benefit most were probably universities which could generate patents but did not have the patent administration and enforcement resources to ensure that users paid up.

The US Patent Office, meanwhile, was snowed under by the several thousand biotechnology-related patent applications it had received. And it was not simply a question of quantity: many of the applications called upon the Patent Office to decide on issues it was not qualified to judge. 'When immunologists cannot agree exactly what constitutes a unique hybridoma,' as one patent attorney put it, 'you can hardly expect patent examiners to do it for them.'

Patenting facilities were coming under fire in a number of countries, in fact. In the UK Margaret Thatcher had recruited her chief scientific adviser, Dr Robin Nicholson, from a leading biotechnology company, Biogen. One of his tasks had been to investigate the country's Patent and Trademark Office. 'Overall,' he concluded, 'the impression given is one of an arcane world, rather than that of modern technological Britain.' Advice given to inventors tended to be 'densely written and full of jargon', while, on the infrequent occasions where the Patent Office took the initiative, it often appeared to have missed the point by emphasising 'clever invention and not its exploitation'. Such comments were given particular spice by the fact that monoclonal antibodies, which as we have seen were invented at the UK's Laboratory of Molecular Biology, were never patented. A Green Paper on intellectual property rights published in 1983 noted that monoclonal antibodies had been a 'free gift to overseas business'.

The already muddied waters were stirred still further by the claims of such companies as Cetus to have developed 'second-generation' versions of such bioproducts as beta interferon. Arguing that they had genetically engineered such products to perform better once inside the human body, they were slapping in further patent applications. The patent examiners will have a hard time deciding whether many of these products are truly novel.

Lawyers in Clover

Meanwhile, however, the industry's patent lawyers were limbering up with some early lawsuits, and looked set to make a great deal more money out of biotechnology than would biotechnologists themselves. Revlon and California's Scripps Clinic & Research Foundation, for example, filed suit against Genentech and Chiron, alleging that these two companies had infringed their patents on a process for purifying Factor VIII, used in the treatment of haemophilia. A major battle was also in the offing over rights to another product of genetic engineering, interleukin-2 (see page 92).

In the diagnostics field, where the goal is to identify the disease which is afflicting a patient, Hybritech, a leading producer of monoclonal antibodies, filed a patent-infringement suit against Monoclonal Antibodies, seen as a competitor in the diagnostic kit field. DNA probes (see page 38) were likewise the source of a good deal of friction, with major competitors including Bethesda Research Laboratories, Cetus and Enzo Biochem. Companies such as Bethesda were marketing diagnostic kits based on technology licensed by Enzo Biochem under a patent applied for but not yet granted.

Clearly, as in a growing number of other cases, this situation injected a good deal of uncertainty into what was already a relatively high-risk market. But perhaps companies like Bethesda took comfort from David Milligan, vice-president for diagnostics R&D at Abbott. 'I would say that no one is going to have a lock on this technology,' he concluded. 'Patents, historically, have not been found to provide the kind of protection in diagnostics that they do in the pharmaceutical area.'

Pharmaceuticals were—and are—where the real action could be found. And no pharmaceutical demonstrated this fact better than alpha interferon (the interferons were among biotechnology's early 'miracle' products, promising a cure for cancer and tremendous profits for those who got them to market first). A key invention, by Dr Charles Weissmann of Zurich University (who became Biogen's chief scientific adviser), had helped transform interferon from a product costing billions of dollars a gram in the 1970s to a substance which Biogen had been literally giving away to doctors and research scientists for clinical trials.

The result of these trials proved less promising than hoped, but alpha interferon still appeared to stop the proliferation of cells in certain rarer blood cancers, and it appeared also to control some viruses and the common cold and to stimulate the body's own defences to attack some tumours. Outsiders watched the developing commercial brawl with relish.

The first alpha-interferon patent went to Biogen. Announcing the impending patent, covering 11 European countries which account for perhaps a third of the world pharmaceutical market, Walter Gilbert said that Biogen and its licensee Schering-Plough 'believe this patent covers the manufacture and sale of any alpha interferon made through recombinant technology'. The patent, which the European Patent Office awarded later in the year, 'gives recognition to Biogen as the first company in the world to make recombinant alpha interferon', he added. But over in California there were dissenting views, to put it mildly.

Indeed, the patent was under attack even before it was published, and a yet more heated exchange was expected when it came to deciding who was to get the key US patent for alpha interferon. The initial challenge came, inevitably, from Genentech and its licensee Hoffmann-La Roche, which argued that the Biogen patent covered only a *precursor* to alpha interferon, a related protein which the body proceeds to develop into alpha interferon. By contrast, it claimed, its own patent application, filed some time after Biogen's, covered alpha interferon proper.

Speaking for Biogen, Dr Weissmann retorted that Genentech was 'nit-picking'. Natural interferon, he said, emerges as a protein with 23 extra amino acids at one end, which are then trimmed off. Biogen's Intron, the subject of the disputed patent, has only seven extra amino acids. Weissmann asserted that Intron was just as effective against viruses and noted that the patent application had been updated to take account of these structural differences. The dispute quickly resolved into two key questions: whose patent application was first, and which covered the *real* interferon.

There was no problem with the first of these questions. Even Genentech admitted that Biogen had filed first in Europe, with its application dated December, 1979, while Genentech's application had gone in seven months later, in July, 1980. Both companies later filed US applications, but under patent law companies can claim earlier European filing dates as the effective date for such US filings. It is not difficult to see why Genentech decided to challenge the Biogen patent, querying whether it was legal to use the earlier patent application 'as a springboard' to cover the final form of alpha interferon.

The Cort Case

The intense drive for patents had already claimed its first public victim: Dr Joseph Cort. A US scientist and one-time Communist who had fled the UK for Czechoslovakia during the 1950s, to escape the McCarthy hearings in the USA, Dr Cort admitted in 1982 that he had fabricated a number of key research results he had reported since returning to the USA in 1976. Following a year-long investigation, the Mount Sinai Medical Center in New York, where Cort had worked, produced some damning conclusions.

The investigating committee announced that Cort had told it that one analogue of the hormone vasopressin which he had reported discovering was fictitious. The committee could find no evidence that another analogue which Cort had claimed had ever been synthesized. It also queried the existence of two variants of the reproductive hormone called 'luteinizing-hormone-releasing hormone' (LHRH), which Cort had reported.

Much of Cort's work had, in fact, been done under contract for Vega Biotechnologies, the company which had been first in the field with commercial 'gene machines' (see page 57). To his horror, Vega president Dr Leon Barstow discovered that a good deal of the research he had bought for $250,000 from Mount

Sinai had more to do with the inventiveness of Cort's imagination than with the inventiveness of his research team.

Cort himself blamed the intense pressure building up in the biotechnology field. 'A lot of molecules had to be made very fast,' he explained in the wake of the scandal. 'You always try to claim as many variations as possible on a patent. I wanted all the molecules to be made that could be made. You're always going ahead as fast as you can—you have to catch up by making molecules that are slightly different.'

No one doubts, however, that the race for patents will go on, not least because companies that intend to stay in the business simply cannot afford to be locked out by someone else's patent. 'We cannot afford to avoid patenting,' said Genex president Dr Leslie Glick, 'but we do it primarily as a defensive measure. Everybody has to do this to protect themselves, but it really is an unproductive use of resources.'

Others agreed. Cetus president Robert Fildes pointed out that most biotechnologists he knew 'would much rather sit down as businessmen than fight in the courts. I think it will be most unfortunate,' he concluded, 'if biotechnology ends up making the patent attorneys wealthy'. One can see his point, but Cetus, which claims to have been the first of the new-breed biotechnology companies, was one of the most striking examples of the way the investor hysteria of the early 1980s turned many *biotechnologists* into millionaires.

THREE
The Midas Touch

'There's always a great deal of excitement in this field,' quipped Cetus chairman Ronald Cape in 1984, 'because of the money involved. Biotechnology is a great way to make a small fortune —as long as you start with a large one!' But, as Cape later pointed out, the biotechnology industry is evolving very rapidly indeed. A few years before, biotechnologists had been talking of a range of hypothetical benefits, he noted, while their critics were worried about very real risks. By 1984, in contrast, biotechnologists were talking about 'very real benefits, while any discussion of risks is hypothetical at best'.

As far as investors were concerned, biotechnology had seemed almost too good to be true. 'Where else,' asked venture capitalist Frederick Adler, 'can you combine making money—which is the primary purpose of venture capital—with the feeling that you're doing some good in the world?'

Setting up a series of venture-capital funds, Adler tended to put a third of the money raised from investors into biotechnology and medicine, with the remaining two-thirds going into electronics companies. 'Biotechnology is moving very rapidly,' he explained, 'but it's going to take much longer to develop. In essence, we can develop companies and bring them to profit in most areas of computer technology in three to four years. There's no such guarantee in biotechnology.'

Interestingly, too, Adler had been involved in one of the first biotechnology companies to get into real financial difficulties, and, indeed, the first biotechnology company to go bankrupt, Armos; he invested $1 million in Armos before deciding, after disagreements with the company's executives, to pull out. Two other companies which had filed for protection under US bankruptcy law, Lee BioMolecular and Southern Biotech, had announced that they would reorganise their businesses to ensure their survival. But, interestingly, the widely expected wave of bankruptcies had failed to materialise.

One of those who perhaps unintentionally fuelled speculation about the financial strength of many of the start-up companies was Robert Johnston, who helped set up such firms as Genex, Cytogen and Ecogen. Speaking to a conference on bio-investment early in 1982, he noted that 'there are approximately 150 independent biotechnology companies in existence and the majority of these, probably 90 per cent, are in the USA. I would estimate that 50 of those companies will not be in existence a year and a half from now. The majority of those 50 will probably go bankrupt as opposed to being acquired.' Even so, he predicted that 'there will be more companies formed, because there is a great need for the technology. However, the people who will form the companies in the future will create them with better concepts and ideas as to what they will do—and how they will do it.'

Some time later, Johnston advised would-be commercial bio-technologists: 'You have got to go out there with a rifle, not a shotgun.' No longer, he said, would any self-respecting venture capitalist 'dream of starting a company as broad as Genex'. Cytogen and Ecogen were much more focused: the former aims to market antibodies for use in treating cancer, while the latter aims to develop new biological pesticides.

Market Guesstimates

Yet, while the stories of the impending 'shake-out' in the biotech-nology industry were certainly premature, the new investment climate symptomatized by such forecasts hit a number of com-panies hard. The vulnerability of much of the industry was forcefully brought home to me when I arrived in California in November, 1982, with a meeting already set up at the Inter-national Plant Research Institute (IPRI). I found the company in crisis—and in no mood to talk. In the wake of a poor stock-market response to a public share issue by Genex, IPRI had found enormous difficulty in raising much-needed resources from pri-vate investors. Unable to pay its November payroll, it started to look around for a company willing to take it over. Ultimately IPRI was taken over by Bio-Rad, but by then it had been forced to slash its staff by over 60 per cent.

Part of the problem had been that hype feeds hype: over-enthusiastic entrepreneurs found no difficulty in coming up with market analysts who would echo their highly optimistic market forecasts. 'Think of a moderately high number,' I wrote in the *Guardian* early in 1982:

Feed it into your calculator and multiply it by a million or, if you are feeling bullish, by an American billion. Dress up the resulting figure . . . in some breathless text and publish. If you can find the nerve to charge several thousand dollars or pounds for the resulting report, so much the better. The market for surveys of the developing market for biotechnology products is itself booming. Yet one conclusion which can have escaped few of those who actually read the results is that the forecasts are so wildly different that only a small minority are likely to be near the mark. The problem is identifying which these might be.

According to market consultants T. A. Sheets & Co., the world market for biotechnology products in the early 1980s was worth around $25 million a year and would grow 2,592-fold by the year 2000, when it was forecast to be worth $64.8 billion. Almost simultaneously, however, another US firm, Business Communications Co., was valuing the current market at $60 million. It forecast that the market would grow to more than $13 billion by 1990. In the UK, IMSWORLD was talking about a figure of $27 billion by 1990. Information Services, meanwhile, were rather more conservative: they valued the then current world market at just $10 million, suggesting that it might grow 50-fold to $500 million by 1990.

Even a mathematical incompetent can see there is a discrepancy between a forecast world market worth $500 million by 1990 and the $27 billion revenues predicted for the same year for the US industry alone. Clearly, different definitions of biotechnology and different assumptions about future growth and market penetration were being used. The overall impression left by such projections, however, was that biotechnology was a Good Thing for investors to be in. The orders went out to buy, buy, buy.

It is perhaps worth looking a little more critically at just one component of one of these studies, that by T. A. Sheets. A couple of months after Sheets went public with his forecasts, Dr Ralph Batchelor of Beecham Pharmaceuticals put the forecasts for antibiotic markets, something Beecham presumably knows a fair amount about, under his microscope. He described the way that Sheets had reached his forecasts as 'facile and stupid'. Sheets had taken a world figure for the current world antibiotics market of $8.5 billion, assumed a 7.5 per cent annual growth rate and ended up with a market worth $33.5 billion in the year 2000. He had then further assumed that biotechnology would have achieved a 10 per cent penetration of the market, producing a figure of $3.3 billion.

The real world, Batchelor pointed out, is very much more complex than one might assume from such statistical juggling. First, he stressed, the pharmaceutical industry had already devoted an enormous effort to improving the efficiency of the strains of bacteria used to produce antibiotics such as penicillin. Indeed, the strains used are highly selected and, as a result, tend to be fairly unstable, reverting back to their less developed, less productive state given the least opportunity. Genetically engineered strains, he suggested, would be even more likely to revert.

Beecham, in fact, had recently had an offer of a costing from a specialist biotechnology company, which proposed to upgrade the performance of an enzyme used in the production of penicillin G. But, Batchelor recalled, 'it just wasn't worth it'. The preliminary figures were so unattractive that they were apparently not even shown to senior management.

Yet, with so much money trying to get into biotechnology, there was no shortage of opportunities for those who were ready to advise investors on where to lay their bets. A growing number of investment firms were employing biotechnology specialists: as a rough rule of thumb, one Wall Street analyst can track about 20 companies, so when the number of publicly held biotechnology companies topped this mark, any investment firm wanting to keep abreast of the field had to have at least one full-time specialist.

But at this stage of the game, even experienced biotechnology hands hit problems. To take just one example of poor timing, the London stockbroking firm of Laing & Cruickshank in 1981 reviewed the performance of biotechnology companies in terms of their return on the capital invested. It then selected a 'top ten' for investors wanting to get into biotechnology: Ajinomoto, Cetus, Davy, Flow General, Fortia, Gist-Brocades, Novo Industri, Ranks Hovis McDougall (RHM), G. D. Searle and Smith-Kline. Analyst Jonathan Allum concluded that Ajinomoto, Gist-Brocades and Novo were the best bets for medium-term returns, but noted that 'in the longer term we favour Cetus as the specialist biotechnology company most likely to succeed'. He also thought that RHM (see page 152) was 'a modestly rated company with an unimpressive record which could be transformed by biotechnology'.

Embarrassingly for Laing & Cruickshank, however, Cetus promptly hit a welter of problems which made it look a great deal less attractive—and Genentech, which had not even been included in the Laing & Cruickshank listing, began to look like the industry leader. When Laing & Cruickshank produced a second portfolio of eleven companies late in 1983, only two of the original companies (Fortia-Pharmacia and Novo) were still recommended. This time, however, Genentech was included.

The Next IBM?

To be fair, the Cetus public offering in 1981 was a phenomenon. It was the second major new biotechnology company to go public, following Genentech's public offering the previous year (page 59), and raised a record $107 million. Its financial results in the previous years had hardly justified this extraordinary largesse on the part of investors. In 1978, it had reported a loss of $1.75 million, in 1979 a loss of $2 million, and in 1980 a small profit. But the timing of the launch was spot-on, with the enthusiasm of the investment community reaching fever pitch. Cetus had proved to have the Midas Touch.

Originally, the three prime movers behind the company were Dr Ronald Cape, a chemist and molecular biologist; Dr Peter Farley, originally a physician; and the Nobel laureate Dr Ronald Glaser, Professor of Physics and Molecular Biology at the University of California, Berkeley.

Peter Farley had often said that he was 'building the next IBM', promising to 'follow the technology wherever it leads'. The trouble was that this strategy spread the company too thin: soon it had 20 major projects under way, without a single major product in its sights. Cetus could—and did—claim to be the richest and most diversified of the biotechnology companies, but it was riding for a fall.

In its early days, before taking that golden shower, Cetus had done most of its work for a subsidiary of Schering-Plough, developing its own proprietary research programme from 1976, following its success in attracting oil-industry investment. Soon, four major companies were funding research programmes at Cetus: National Distillers and Chemical (interested in ethanol production); Shell Oil (interferons); Standard Oil of California, better known as Socal (fructose and alkene oxides); and Standard Oil of Indiana (hydrocarbon modification).

The key programme was almost certainly that targeted on the conversion of glucose to fructose, a high-value sugar found in honey and many fruits, with alkene oxides as a by-product. Fructose was an unusual target for a new biotechnology company to pick because it is a commodity product. As Cape himself had said, fructose is 'one of the large, commodity items where if you can't come in at the right price you don't have any market at all'. By contrast, many companies were concentrating on small, specialist niches where they could out-manoeuvre larger companies and, to some extent, dictate the price of their product.

'The question is,' Cape had said, 'can we deliver in sufficient quantity at the right price in order to interest people like Coca Cola and Pepsi Cola, who between them buy 25 per cent of all sugar bought in the United States? The problem is that the capital costs in fructose production are enormous. It will take six or seven years to commercialise, at a cost to Cetus and Socal of some $150 million.' The potential market, however, was very attractive indeed. Cetus expected to win perhaps 20–40 per cent of a market valued, in 1982, at $10 billion in the USA alone. And then, horror of horrors, Socal pulled out of the project.

Like other once-buoyant companies, Cetus was soon forced to cut back its staff from a peak figure of over 540. It also beefed up its top management by poaching Dr Robert Fildes from Biogen as its new president and chief operating officer. Fildes, who had been president of Biogen and, prior to that, vice-president of the industrial division at Bristol Myers, was the only person to have served as president of two of the largest four US biotechnology companies. And, before long, Farley was out.

'Dr Farley, who served for five years as Cetus' president, is an industry pioneer,' said Cape, announcing Farley's resignation. 'Pete helped to create an industry as well as a leading biotechnology company.' The change, in fact, was symptomatic of the fundamental restructuring that was then under way not only in Cetus but in the whole biotechnology industry. Maturing companies tend to shake out some of the entrepreneurial talent that got them going in the first place, just as Dr Martin Apple had been shaken out of IPRI, some time before the crunch came (see pages 104–106).

Cetus was trying to focus its activities. Shortly after its first stock offering, it had given probable time-scales for some of its products:

veterinary vaccines (time to market, 1–1½ years)
diagnostic products (2½ years)
agribusiness applications (3 years)
interferons (4 years)
fructose (6 years, which helps explain Socal's decision)
chemicals and energy derived from plant material or
    biomass (10 years)

In its refocusing, Cetus concentrated on just three main areas:

diagnostic products, cancer therapies and agriculture. (Its longer-term link-up with Weyerhaeueser on forest biotechnology is described in Chapter 6, pages 114–15.)

Soon, too, it got its first real products on the market, a diagnostic test for cytomegalovirus, for use in blood-transfusion centres, and a vaccine called *LitterGuard* to protect pigs against a diarrhoeal disease called scours. But the competition in this field was indicated by the fact that a rival vaccine was launched almost simultaneously by Salsbury Laboratories, while nine months previously a Dutch firm, Intervet International, had beaten everyone to market with its recombinant scours vaccine. And, just in case anyone imagines that vaccines were the only answer, Molecular Genetics was also soon selling a monoclonal antibody (page 133) which proved highly successful against the same disease.

The Bubble Bursts

By the time some of the second wave of start-up companies began to come to Wall Street, they met a much cooler reception, with Amgen considered the company with the unhappy distinction of having burst the biotechnology bubble. Many companies, including Advanced Genetic Sciences, Biotechnology General, Chiron and Integrated Genetics, were forced to scale down their public offerings in terms of the number of shares, their price, or both.

Even so, many countries still viewed the US venture-capital situation with unconcealed envy. The US venture-capital industry is vast, with Stanford University and MIT representing twin peaks in the landscape. While many universities have preferred to deal with existing companies, MIT and Stanford have consistently tried to 'lower the threshold' between academics and industry, encouraging new start-up companies. Both Boston, just outside which MIT is based, and California enjoy large pools of venture capital. Some of the more successful venture-capital outfits claim portfolio returns of 50 per cent *a year* over seven years, with the average close to 20 per cent.

Such venture-capital operations typically expect to write off about 25 per cent of the companies they invest in, as Adler did in the case of Armos. They hope that about half will show a return of up to 50 per cent on their investment, and that the remaining 25 per cent will produce an even higher return.

To ensure that they get the pick of the crop, they need to be highly selective in their choice of investment vehicles. Of the average 100 proposals presented to them, perhaps 50 will be dropped on first sight and another 40 after further study, leaving 10 for negotiation. Of these 4 will prove unsuitable, 2 will be suitable but out of reach, and 4 will be picked up and funded. To take a UK example, Prutec, set up in 1980 by Prudential Insurance, had talked to some 4,000 people wanting money by 1984. It took about 100 of these supplicants seriously, and actually offered money to just 25. Prutec chief executive Dr Derek Allam admitted disappointment. 'If quality had been there,' he said, 'Prutec would have invested more. We are not constrained by lack of money.'

Anyone who still needs convincing that the US investment scene is radically different should consider the experience of the first UK biotechnology company to go public, Cambridge Life Sciences—which, incidentally, made it into Laing & Cruickshank's second 'top ten'. The company's joint managing directors, Drs Michael Gronow and William McCrae, had previously worked for Patscentre International, the leading science and technology consultancy, helping to establish its biotechnology activities. Their plans for their new company were fairly straightforward. 'We originally had the quite modest intention of setting up a garage operation to extract enzymes from animal bits and pieces,' explained McCrae, not entirely in jest. 'We chose enzymes because of their known commercial value and the fact that the vast majority used in this country are, at great expense, imported.'

Ultimately, however, they concluded that they would require very much larger sums than they had budgeted for. The original idea had been to build up the business slowly, using consultancy projects to generate short-term income. But reality intervened. 'Up to this point,' McCrae recalls, 'we had made the mistake of not discussing our plans with our wives. When they finally heard our intentions for the garages, and in particular what we proposed to keep in their freezers, they used their power of veto —and that was that.'

So Gronow and McCrae began approaching City institutions, only to find that the commercial potential of biotechnology was still far from obvious to those institutions. This was 1980, the year that saw the publication of the Spinks Report on biotechnology— a copy of which they duly attached to their proposal, which was, they believed, 'a formidable and irresistible document'. It was submitted to banks, investment houses and venture-capital organisations, but without success.

Forced to rethink, they concluded that 'our proposal, indeed our whole approach to obtaining finance, had been based on the US model. We had noted how they had set up and developed their biotechnology companies and it seemed a workable and sensible format. We came to realise, however, that the difference in attitude towards venture capital in the UK and the US was greater than we had thought.' In the end, they got money from the Industrial and Commercial Finance Corporation, later called Investors in Industry or '3i'. ICFC suggested that CLS broaden its management base, which was soon done, and identify a product which had a quantifiable market and a time-scale for profitability. This proved much harder.

Gronow and McCrae retorted that 'biotechnology was about future products', that it was 'an area of science where the products would create their own markets'. But he who pays the piper calls the tune, so they picked urokinase as their target product. This enzyme is used to treat thrombosis and related diseases and has a small UK market overshadowed by a Japanese market which at the time was worth over $150 million a year.

CLS expected to be able to start selling urokinase by early 1983, with minimal problems from the regulatory authorities, since it is a product derived from the human body itself, has a low toxicity and does not stimulate antibody formation or allergies. However, the appearance of cut-price urokinase on the European market shortly after CLS launched its own product effectively scuppered its plans. The work was 'put on the back-burner'. A useful spin-off from this work, however, was the development of a novel cell-culture method for producing both urokinase and tissue plasminogen activator—which is even more effective in the treatment of blood clots than urokinase.

Later, CLS raised £900,000, largely from high-bracket-tax investors who, under the terms of the UK's 1981 Finance Act, could make an investment of up to £10,000 in start-up companies in a single tax year and set this against income tax. Late in 1983, the company raised a further £1.5 million from institutional investors, allowing it to expand into the field of enzyme diagnostics, which had long been in the minds of its founders.

CLS plans also to become a centre of excellence in the field of biosensors (see page 213), which blend microelectronics and biochemistry. Contemporary bioelectronics is simply an extension of the diagnostics field, but ultimately it is hoped that it will include miniaturised monitoring systems which can actually be implanted in the human body, revolutionising the whole area of clinical analysis and diagnosis.

Another company which illustrates some of the difficulties involved in getting a small start-up off the ground is Bio-Isolates. Perhaps more than any other company, Bio-Isolates benefited from the bio-euphoria which swept the investment community in the early 1980s. Launched in 1982 on the Unlisted Securities Market (USM) at 33p each, the company's shares were soon trading at a phenomenal £3.50 apiece. No one was more surprised than the company's own directors, two of whom suddenly found their personal holdings worth almost £2 million each. The irony was that those same two directors, Douglas Palmer (now chairman of Bio-Isolates) and Dr Rod Dove, had been made redundant in 1977 by Viscose Development, part of the French multinational Spontex, which decided that the protein-extraction process they had been working on, and which later became the core technology at Bio-Isolates, was not a commercial proposition.

To be fair to Spontex, Palmer and Dove had also failed to attract funds on suitable terms from the assortment of venture-capital firms they approached prior to their USM launch. The first real breakthrough, once Bio-Isolates had bought the patents on the basic protein-extraction technology, came in 1978 when the Welsh Development Agency lent it £100,000 to get started. The Agency recouped its investment—and made a £200,000 profit—when it sold 500,000 of the company's shares in late 1982 at an average price of 60p.

But why the extraordinary enthusiasm for Bio-Isolates? 'We have a product which is right for the market and which embraces the romance of biotechnology,' Palmer suggested several months after the USM launch, 'plus the sale of shares happened at a time when the Stock Exchange was down and needed a shot in the arm.' And a couple of months later still, with the share price still rising, he mused that 'the Government's statement that they meant to invest in biotechnology may have something to do with it. Possibly we are in the right place at the right time, with a product just about ready to become commercial.' This product was an unusual one for a new biotechnology start-up, because it was a food product for human consumption (see page 142).

Meanwhile, whatever Bio-Isolates' particular strengths may have been, the very fact that it had a USM listing brought a number of problems. The majority of the companies which have gone public via the USM believe that the advantages outweigh the disadvantages—in fact, by 1984 the USM had created over 280 'paper millionaires', assessed in terms of their unrealised holdings in USM companies. But the disadvantage mentioned most often was that USM companies become the subject of increased public interest and scrutiny. They also become vulnerable to rumour.

In the case of Bio-Isolates, a variety of early rumours boosted its shares, but also caused some damage to the company. By the beginning of 1983, its shares were valued at £4.50, although they later slipped to around £1.40. According to press reports appearing shortly after the USM launch, none of which originated from Bio-Isolates, the company was rumoured to be on the brink of signing a collaborative agreement with SmithKline, the US pharmaceuticals company (no such agreement, in fact, was contemplated); on the verge of linking up with the US hamburger giant MacDonald's (no such agreement); and about to be taken over by RHM (no such take-over was ever considered).

Other rumours, however, proved better founded. It was rumoured, for example, that Bio-Isolates was about to get a US share quotation and the company's shares did in fact go on sale over-the-counter there early in 1983, lifting them from £2.85 to £4.25 in short order. And it was able to announce real joint ventures, including one with Ireland's largest agricultural co-operative, the Mitchelstown Co-operative Agricultural Society.

The stories of Bio-Isolates and CLS illustrate another key dimension of the biotechnology revolution: they both had close working relationships with university researchers. Bio-Isolates had very close links with the biochemical engineering research group at University College, Swansea, while CLS had links with a number of university biotechnologists.

A Window on the Action

Based in Cambridge, CLS hoped for increasingly close contacts with Dr Christopher Lowe, who had just taken over as the director of the Cambridge Biotechnology Centre—designed to help commercialise some of the exciting biotechnology work being done around Cambridge University. Despite the controversy which had raged in many countries about ever-closer relationships between university biotechnologists and industry, Lowe had no doubts about the need for stronger industry–university ties. I asked him why he had taken on the Cambridge Biotechnology Centre. He explained:

Cambridge offered me the opportunity I wanted, which was a fair degree of independence, coupled with the ability to set up a new venture from scratch. Cambridge is a world-renowned university with tremendous resources and considerable expertise in areas related to biotechnology, while the surrounding region is becoming one of Europe's key concentrations of high-technology industries. The pre-eminence of Cambridge in molecular, cell and plant biology and in chemical engineering is undisputed. The University's research has been absolutely fundamental to the development of biotechnology as we know it today.

So much for the PR, but what was the Centre actually planning to do?

Initially housed in a totally refurbished building on the Downing site, the Centre was adjacent to the departments of biochemistry, botany, chemical engineering, genetics and pathology—but Lowe pointed out that it would draw on a broader diversity of expertise in and around Cambridge. He stressed:

We have a very wide range of interests, if you go from our interests in plant biology and plant cells right through to our work on biosensors and bioelectronics. We should eventually be interacting with the botanists, zoologists, biochemists, geneticists, engineers, chemists, electrical and physical chemists, chemical engineers and the electronics department.

Lowe had strong views on the biosensor prospect (see page 50).

I think that a lot of people have been exaggerating this whole area. There are a lot of problems associated with these things. We have a lot of devices under development, with some through to development stage. Whether we shall take them any further is dependent on the extent of commercial interest. Every device we look at has its problems which have to be resolved and they're not the sort of problems which can be resolved easily. They take time.

And money.

Lowe was discussing joint projects with a large number of companies, and was finding a good deal of real interest, but he felt that it was extremely important to pick the *right* companies to work with.

I'm looking for companies which can offer us more than just money, companies which can give us some technology input that's going to stand us in good stead in some 10 to 15 years' time. There's no point in going after lucrative projects which drain your technical resources. So we hope to link with companies which are firmly in the high-technology game—so that in 20 years' time we are still number one.

Of course, with Cambridge's resources and reputation behind him, Lowe could afford to be a little fussy, a luxury which cannot be afforded by many university departments trying to make ends meet with whatever industrial contract money they can get. But even the Cambridge Biotechnology Centre had hit snags. Like other entrepreneurial academics, Lowe was critical of the performance of much of British industry.

The UK leads the world in biosensors at the moment. I think that's undisputed. But many UK companies are simply sitting on the sidelines, keeping what they call 'a window on the action'. If the position does not change soon, we shall almost certainly be collaborating with foreign companies on research which could have been of direct commercial value to UK companies. It's tragic.

Attracting the services of such university resources, often in the shape of key individual scientists, has been a number-one priority for most start-up companies. Those companies which have managed, like Biogen, Cetus or Genentech, to attract Nobel laureates know what they are doing when they publicise the fact: investors want to see some real scientific weight behind a company, even if the critical work is often going to be done by much younger (and less expensive) scientists.

But science, although important, is only one ingredient in a successful start-up: just as important is the calibre of management and, later, of a company's marketing. 'We must be the biggest unpaid head-hunting organisation in this country,' said Advent managing director David Cooksey, stressing that 'people are the key'. Advent, which at the time managed two high-technology investment funds, had put fully half of its efforts into identifying and acquiring the right people to ensure the rapid, sustained growth of the companies in which it had invested.

A Rothschild Invests in a Revolution

Meanwhile, as the only investment company set up specifically to invest in quoted and unquoted biotechnology stocks, Biotechnology Investments Limited (BIL) had not unnaturally attracted a good deal of interest. With Lord Rothschild and the late Dr Alfred Spinks, chairman of the committee which produced the influential Spinks Report, among its founder directors, BIL is unquestionably an élite operation. It invests in the shares of other biotechnology companies, with its own shares quoted on the Luxembourg and London exchanges.

The company, explained Lord Rothschild, 'was established to allow investors to participate in one of the major technological revolutions of the twentieth century'. Biotechnology, he was convinced, 'has enormous potential and its future impact could well match that of transistors, computers and microprocessors'. But, he stressed, anyone investing in BIL should regard that investment as 'long-term, between three and seven years, and maybe longer. This may seem a distant objective,' he admitted, 'but investment in this exciting field should be approached with both patience and realism.'

BIL's long-term objective has been to develop a diversified portfolio covering companies involved in the application of bio-technology to the pharmaceuticals, chemicals, energy, pollution control, mining and agriculture industries. Its quoted investments were in such companies as Ajinomoto, Alfa-Laval, Flow General, Genentech, Genex, Gist-Brocades, Kyowa Hakko, Millipore, Novo and Pharmacia.

As far as its unquoted investments were concerned, BIL adopted five key criteria. First, the company under consideration should not only employ high-calibre scientists but should also have first-rate business managers able to establish and maintain a successful venture. Second, the company's business plan should clearly define its research and product areas, having both a market analysis and a realistic assessment of the competition. Third, the projected return must be appropriate for the risk taken: BIL looks for 'a much higher than average return' on the unquoted element of its portfolio. Fourth, the company should have a sponsor to act as lead investor, accepting responsibility for the venture, a role which BIL will sometimes perform. And, fifth, there must be a prospective method of realising the investment, typically through a public share issue.

Its early unquoted investments went into such companies as Agrigenetics, Amgen, Applied Biosystems, Catalytica, DNA Plant Technology, Genetic Systems, Immunex, Integrated Genetics and Repligen. Of these, Amgen, Applied Biosystems, Integrated Genetics and Immunex all went public in the 1982–83 fiscal year. 'The rapid transition of these companies from start-up investments to the public market is encouraging,' said Lord Rothschild, but he cautioned that 'this has been possible because of the recent strength of Wall Street. It should not be assumed that this time-scale will normally apply to the public offering of BIL's investments.'

To begin with, BIL could find no UK investment opportunities which met its rigorous criteria, although it did finally invest in

Celltech. Its next wave of investments was, once again, in US companies, such as Genzyme, Plant Genetics and Queue Systems. Soon, however, it was making investments also in such UK companies as IQ(Bio), Twyford Plant Laboratories and WMC Resource Recovery. Meanwhile, its stakes in companies like Flow General, Pharmacia and Queue Systems were an interesting indicator of a significant new trend in biotechnology investing. Said Peter Laing of BIL during the uncertain days of 1982:

> Investment in biotechnology is not very different from any other investment. It tends to be cyclical and a number of discernible patterns can be seen. A project recently offered to us on the production of electricity from algae reminded me of that intense period of speculation in the eighteenth century in a number of UK companies that were set up to exploit the new discoveries in the Pacific and the South Atlantic. This period was known as the South Sea Bubble, and one of the last companies to come to the market before the bubble collapsed and people lost all their money was one that was set up to extract sunlight from cucumbers.

Selling Picks and Shovels

So far, Laing admitted, he had not been offered 'a bacterium which will make gold, although immediately after the First World War the German Government, which was faced with immense war debts, did try to extract gold from sea water using bacteria'. And then he went on to paint in the background to a key emerging trend in biotechnology investment: a growing interest in 'picks and shovels'. The people who made a lot of money out of the Klondike gold rush, he pointed out, were not those who rushed in to find gold: instead, they were the people who sold picks, shovels and tents to those who were digging for gold.

With some 10,000 private and public laboratories estimated to be involved in biotechnology worldwide, there has been a growing appetite for products such as refrigerators, centrifuges and the glassware used in such work. We have learned to be suspicious of market forecasts, but Business Communication Co. projected in 1984 that, from a 1983 base figure of $1.6 billion, sales of the chemicals, equipment and instrumentation used by the US biotechnology industry alone would grow at an annual rate of 13 per cent through to 1990, reaching a value of $3.3 billion that year. By 1996, BCC predicted, the growth rate should still be around 13 per cent, and the market should be worth $6.2 billion.

Gene machines, which string together synthetic DNA building blocks into customised gene sequences, are at the more sophisticated end of the pick-and-shovel spectrum. 'By automating and thereby speeding the gene-synthesis process,' said one analyst of such machines, 'they are providing researchers with ample quantities of DNA, allowing them to spend more time on research and less on gene synthesis, and greatly increasing their chances of finding commercially viable compounds.' At least six US companies are making gene machines, including Applied Biosystems, Biosearch, Genetic Design Sequemat, Systec, Vega Biotechnologies and SmithKline Beckman. Several European companies are also involved in this market, including Celltech, West Germany's Biochemische Synthesetechnik and Sweden's Analysteknik.

But there have been some problems in this new field. Vega Biotechnologies (see page 40) was first to the market with a gene machine, but later had to recall and upgrade those it had sold —at a reported cost of $4,000 apiece. Coupled with their high price, such problems helped depress the demand for DNA synthesizers. Bio Logicals had to drop out completely, after spending $3 million on gene-machine R&D.

Applied Biosystems, which I visited late in 1982, did much better. Located on an industrial estate in Foster City, on San Francisco Bay, Applied Biosystems really had begun life in a garage. By late 1983, however, the company was showing a healthy profit. 'Revenues have grown faster than we can efficiently spend on R&D,' said Dr Sam Eletr, the company's president. Systec, based in Minneapolis, was also doing well, although company president Richard Zelinka said he was trying to keep his company small until its roots were secure. 'We've got venture-capital people crawling all over us,' he told the *Wall Street Journal*. 'We tell 'em we're not ready yet.'

A longer-term problem with this market, however, is that it is limited: it has been estimated that there are only about 7,000 potential industrial and university users. So these companies will either have to diversify or go out of business. As the *Journal* noted, 'one way they're doing that is by imitating Gillette. Makers of shaving supplies learned that the real fortune came in consumable blades, not in the one-time purchase of hardware. So companies are supplementing gene-machine sales with sales of the chemical ingredients, or reagents, that make DNA.'

Very few biotechnology companies have been truly profitable so far, however. But by 1984 an increasing number were beginning to turn the corner. Cetus, for example, had turned a loss of $5.7 million in fiscal 1983 into a net income of $992,800 in fiscal

1984—despite a 35 per cent increase in costs, largely due to staff increases and product-testing programmes. It also sold 51 per cent of its agricultural subsidiary, Cetus Madison, to W. R. Grace, which was talking of ploughing at least $60 million into the resulting agricultural biotechnology joint-venture company, called Agracetus.

Genex was just one of the services-oriented companies which continued their transformation into product-based companies. Although Genex reported a $5.4 million loss on 1983 revenues of $11 million, it was one of the best examples of a company which, having established its base by teaching others to make money from biotechnology, began to make real biotechnology money itself.

'It's Basically Free Money'

One sure-fire prediction we can make is that the new biotechnology companies, which have not been noted for any modesty in their demands for investors' money, will become even more money-hungry as they begin to take their processes from the laboratory bench to full-scale production. The $2.5 billion invested by 1984 was only the beginning.

There has been a succession of new ideas for raising money, including the 'R&D limited partnerships' whose attractions led one venture capitalist to exclaim: 'It's basically free money.' The willingness of wealthy investors to put their money into high-risk ventures has always been a marked feature of the US scene, and such partnerships were simply the latest tax-shelter scheme. Their attractions include the scale of resources they can harness: Agrigenetics was the first biotechnology company to raise money this way, with a $55 million partnership in 1981. Cetus raised $75 million in a single deal, while Genentech took $90 million in two.

Such partnerships enable investors to put their money into specific projects, rather than into a company's equity. The 'general partner', which may be the company, a subsidiary or some third party, manages the partnership, while the investors (or 'limited partners') can deduct money invested from their income —often taxed at the 50 per cent level. Any income is taxed only at the lower capital-gains level of 20 per cent. The advantages for the general partner include the fact that the money raised shows up in financial statements not as a debt but as contract revenues, enabling it to raise more money.

But there is no such thing as 'free money'—or, at least, precious little of it. If the research fails, some lawyers expect that the general partner will be inundated with lawsuits by disappointed investors. Also, to show the right sort of profit, the partnership's products must be extremely successful. A $75 million partnership promising investors a 5-to-1 return, for example, will ultimately have to pay out $375 million. As the journal *Bio/Technology* pointed out, 'since the money is derived from royalties of 6 to 10 percent, the partnership must achieve well over $4 billion in sales over the next 15–20 years to make good its claims. And these,' it noted, 'are predominantly wholesale rather than retail sales.'

After the initial enthusiasm, a weakening in demand from investors for such partnership interests was blamed for Biogen's failure to raise the full $40 million it was looking for in 1984 to fund the commercial development of its gamma-interferon and interleukin-2 products. Biogen raised just $25 million, and so will have to invest $35 million in the $60 million partnership, rather than the $20 million it originally budgeted. Centocor was another biotechnology company which hit such problems. It tried to raise up to $15 million for its work on cancer genes (oncogenes) but later said that it would be happy to settle for $5–10 million.

But a number of leading biotechnology companies continued to predict that they would soon achieve a $1 billion value. Genentech, said the company's president, the venture capitalist Robert Swanson, 'will be a $1 billion company by 1990'. Started in 1976 by Swanson and Stanford University's Herbert Boyer, who put up $500 apiece to form the company, Genentech aims to be the first start-up since Syntex to become a major pharmaceutical company. It has regularly trumped the competition with its announcements of key breakthroughs. The first company to splice a human gene (for a brain hormone, somatostatin) into a bacterial cell, Genentech was also the first major biotechnology company to go public. October 14th, 1980, will long be remembered on Wall Street. Genentech went public that day, and shares issued for $35 apiece at 10am hit $86 by lunchtime. The company's prospectus had warned, however, that its common stock 'may be highly volatile' and, before long, its share prices had fallen close to the original offer price.

From the very beginning, Genentech set its sights on products where the potential returns 'are measured in dollars per milligram instead of cents per pound', which has inevitably meant pharmaceuticals. It also emphasised 'critical mass' in building up its scientific resources. The ability to assemble multi-disciplinary teams in short order, and generally from under the same corporate roof, has proved to be a key factor in the company's continuing success. Indeed, as James Watson pointed out in *The Double Helix* (1968), one of the main reasons he and Crick beat Maurice Wilkins to the establishment of the structure of DNA was that Wilkins worked 'in a lab devoid of structural chemists' and, as a result, 'did not have anyone to tell him that the textbook pictures [of DNA] were wrong'.

A prime example of what 'critical mass' can do for you. The problem has been, however, that the 'critical mass' of scientists, engineers and other resources you need to break into any given field of biotechnology has tended to increase sharply as the easier targets have been mopped up. Some of the new equipment, such as DNA synthesizers, can help cut the number of people you need to employ; but still the rising 'entry fee' is one of the reasons why venture capitalists like Robert Johnston are setting up Cytogens and Ecogens rather than more Genexes. To get an idea of what the world looks like from inside the gene factory, and how existing product ranges are coming under pressure, we'll examine two of the leading companies operating in biotechnology: Biogen and Amersham.

FOUR
Inside the Gene Factory

'I may sound like an Arab sheikh with ten wives who always says that the last one I married was best,' said the venture capitalist Moshe Alafi in 1980, 'but this is the best company I've been associated with.' He was talking about Biogen and, having been involved in the launch of a dozen or so high-tech companies in the South San Francisco Bay area, he presumably knew what he was talking about. Biogen, in fact, has been unique in a number of ways but, as it is a leading company in the biotechnology business, its brief history and current targets tell us a good deal about the way the business as a whole has been developing.

If any one man can take the credit for catalysing the process which led to the launch of Biogen it was Ray Schaefer, a former vice-president for research and development at the US subsidiary of Inco Ltd, the huge Canadian nickel producer. In 1976, Schaefer was scouting around for investment opportunities for Inco, which wanted to diversify into new business areas, when he came across Cetus. Alafi, who had helped set Cetus up, was then chairman of the company and encouraged Schaefer to make a small investment. Schaefer bought his Cetus stock from another San Francisco venture capitalist, Peter Perkins, who was also helping get Genentech off the ground. As a result, Inco also bought 15 per cent of Genentech's stock.

But that was just the beginning for Schaefer, who by now had been bitten by the biotechnology bug. Struck by the fact that Cetus had been able to sign up two US Nobel Prize winners, he wondered whether European laureates could be recruited in the same way. Inco was supportive, so Schaefer toured Europe with Dan Adams, who headed Inco's own venture-capital group. Most of the people they visited first time around were heads of university molecular-biology departments. They met with a frosty reception.

Many such scientists had operated as consultants from time to time, but they were acutely concerned that agreeing to a permanent link with a biotechnology company like Biogen could damage their professional standing. Schaefer and Adams kept plugging away, however, and there was no denying the attractions of the financial package they were offering—including general research grants to compensate both the scientists and their universities for any time spent working with the new company.

In the end, ten scientists signed up, three American, seven European. In at least one case, the decision to join Biogen caused considerable controversy. This was in the UK, where growing concern about the biotechnology 'brain drain' found a focus in the news that Dr Richard Flavell would join Biogen. Flavell had headed a laboratory investigating gene structure and expression at the National Institute for Medical Research (NIMR), based in Mill Hill, London. Although he officially took 'leave of absence', it required little imagination to predict that he was unlikely to return to NIMR at the end of his two-year 'secondment' to Biogen's new laboratory in Cambridge, Massachusetts.

Biogen was formally launched in May 1978. It began life with a sum of $750,000, which came from Inco, venture capitalists T. A. Associates, Moshe Alafi and a number of European investors. It promptly began to plough some of the money into research programmes run by some of its new scientific figureheads. It backed work which Professor Charles Weissman and his team at the University of Zurich were carrying out on interferon. It helped underwrite Professor Walter Gilbert's studies at Harvard on the bacterial production of insulin. And it supported hepatitis research at the universities of Edinburgh and Heidelberg.

The Edinburgh work was being carried out by Professor Kenneth Murray, whose work (in collaboration with his wife Noreen) on the bacteriophage lambda (a virus) had resulted in the development of phage cloning vectors—molecules which carry recombinant genes into host bacteria, natural genetic engineers which are now being developed as therapeutic tools (see page 101). Murray had also succeeded in cloning and expressing hepatitis virus DNA, a key step towards the development of a hepatitis vaccine.

Hands on the Jugular

Many of the new biotechnology companies had recruited eminent scientists to their advisory boards, but Biogen believed that its approach was different. 'This is the only company in the world,' Moshe Alafi explained, 'where scientists have their hands on the company's jugular vein.' In effect, it had set up what it once described as a 'mini-academy', which met frequently, often at weekends.

'The long-range prospects are for sales of billions of dollars' worth of products,' Schaefer predicted, 'but the idea that anyone will make money right away is crazy.' Well, not entirely crazy. Biogen's key scientists were soon scientist-millionaires, even if only on paper: by mid-1980 nine of the ten were worth $1.5 million apiece. Weissman, whose success in genetically engineering bacteria to produce a version of human alpha interferon had captured the world's headlines earlier that year, was rewarded with a further block of Biogen stock; significantly, he had been working on mouse interferon until his fellow shareholders in Biogen suggested that human interferon might be a rather more attractive commercial target. 'Critical mass' again.

All was not well on the business side of the operation, however. After a good deal of searching, the company finally found a full-time president in November, 1979. Robert Cawthorn had previously been executive vice-president with Pfizer-Europe. Adams, who had helped run Biogen as general manager, had left after a set-to with Inco—and neither Schaefer nor Alafi were prepared to take on the job. In the event, Adams was replaced by Wally Gilbert, chairman of the company's scientific board, who took on the job of acting president. Gilbert, who was the American Cancer Society Professor of Molecular Biology at Harvard, had won the Nobel Prize for Chemistry in 1980. His importance to the company can be judged from the fact that early on Biogen took out a 'key-man' insurance policy on his life to the value of $5 million.

Things seemed to be looking up for Biogen. Although it had difficulty raising further investment in Europe, where biotechnology was still seen as a particularly high-risk business, it persuaded Schering-Plough to invest $8 million in return for worldwide licensing rights on three Biogen products of its choice, while Inco chipped in another $1.25 million. Biogen had identified a number of 'strategic directions': its main areas of interest included pharmaceuticals, chemicals, energy, agriculture, food, mining (focusing on microbial metal-leaching: see page 220) and pollution control—or, to put it another way, Biogen was backing most of the horses in the race. In part, this resulted from the widely different business interests of its major corporate sponsors: Inco (mining and metals refining), Schering-Plough (pharmaceuticals), Monsanto (chemicals, agriculture and energy), and Grand Metropolitan (food and beverages).

Later, in 1982, Biogen linked up with Stone & Webster, the

US engineering contractor, one of a number of engineering firms which formed joint ventures with biotechnology companies to capture a share of the expected new business as the biotechnology industry began to scale-up its operations. As far as Stone & Webster was concerned, a key interest was Biogen's ethanol-from-biomass technology, developed by Professor Daniel Wang of MIT.

But Biogen's centre of gravity was always in the pharmaceutical field. One of its first major announcements, for example, had come early in 1980, when it reported that Weissman's team at the University of Zurich had successfully inserted a human interferon gene into E. *coli* and got the bacteria to churn out interferon (see page 90). The only fly in the ointment—which Biogen did not publicise—was the fact that the resulting bacterial interferon was not identical to human interferon; bacteria, unlike human cells, cannot attach sugar chains to complex proteins.

Cawthorn, however, was delighted, saying that the achievement had 'put Biogen on the map'. Competitors grumbled that Biogen had held its press conference at a ridiculously early stage in the product's development, going public 'with a crude method to get the jump, for commercial reasons, on other teams doing the same work'. But this was perhaps a case of sour grapes. While Biogen's initial process could produce only one-thousandth of the amount of interferon produced by human cells, the new companies were already manoeuvring for commercial advantage —and Biogen had beaten its competitors to the draw, at least on the publicity front. This was simply the first skirmish in what later developed into a full-scale battle between Biogen and Genentech over the patenting of alpha interferon.

'The Race Is On'

'The race is on to reach the market with socially desirable and commercially acceptable products,' said Dr Gilbert. But, he admitted, 'selecting new pharmaceutical products is very difficult. Our strategy is based on the conviction that there will be a large number of bioactive human proteins that can, using genetic engineering, be made easily and cheaply in bacteria.' He was by no means convinced, however, that genetic engineering would be a universal panacea: he pointed out that Biogen was working for Novo Industri on genetically engineered insulin, but noted that Novo's chemically treated insulin derived from pig pancreases could well retain its commercial lead.

Such joint ventures with larger companies have been a key

tactic in the biotechnology business, helping attract the necessary resources—and spread the risks. Biogen has signed a wide variety of agreements with other companies. Schering-Plough, for example, which had invested in Biogen on the basis that it had first choice on three unspecified products, won worldwide licensing rights for Biogen's alpha interferon, beta interferon and erythropoietin (used in the treatment of kidney disease). Inco got worldwide rights on Biogen's bovine and swine growth hormones, while four companies got non-exclusive rights to Biogen's first actual product, a spin-off from the work on hepatitis vaccines, a kit for diagnosing hepatitis; the four successful companies were Behringwerke, Green Cross, Hoffmann-La Roche and Sorin.

Biogen has also actively pursued links with Japanese companies, and broke new ground when it signed an agreement, late in 1983, with China's Shaanxi Pharmaceutical Bureau, focusing on the development and marketing of gamma interferon. The agreement reflected China's growing interest in both genetic engineering and cancer therapies. Over a million new cases each of stomach and liver cancer are reported every year in the world's most populous country.

Although diagnostic kits should bring in some useful income and establish Biogen's presence in the marketplace, Gilbert sees the recombinant-DNA products as the real potential money-spinners. A particular attraction of diagnostic kits, however, is that they do not have to pass through all the clinical trials to which pharmaceuticals are subjected—meaning that they can be brought to market very much faster.

Even when a clinical trial is under way it can still come up with problematic results, undermining investor confidence. For example, shortly before Biogen went public on Wall Street, trials of its genetically engineered alpha interferon hit problems. A nasal interferon spray was being tested as a possible cure for the common cold in both the US and the UK but, worryingly, it was found that results obtained in human volunteers in the special conditions of a colds research unit could not be reproduced in people who had caught colds naturally. Worse, the doses of interferon seemed too high: some of those treated seemed to be showing cold-like symptoms due to the interferon itself.

Partly in order to finance its clinical trials of such products as interleukin-2 (in cancer and AIDS patients) and gamma interferon (in cancer patients), Biogen decided the time had come to go public. By the time it launched itself on the stock market in 1983, Biogen had raised something like $80 million from its corporate sponsors, but needed more money to mount an effective challenge to such cash-rich competitors as Cetus and Genentech. The cost of Biogen's expansion plans and clinical trials showed up in its results for 1983: although it reported revenues of $18.4 million, it had also run up expenses of $30.1 million.

In contrast to the mood when companies such as Genentech and Cetus went public, however, investors were a good deal more cautious about biotechnology stock issues by the time Biogen produced its first prospectus for public consumption. Even so, it succeeded in raising $57 million. Indeed, its offering had been postponed for some months, for fear that investor reaction would be embarrassingly cool. 'Interest should never have been as high as it was,' said Alan Burg, an analyst with Arthur D. Little & Co., of the Genentech phase of the biotechnology bubble, 'and it never should have declined so dramatically.' Biogen had suffered a series of management shake-ups, however, which had not helped its image with investors. Apart from management problems, analysts could point to a number of other possible indicators of weakness: genetically engineered insulin produced by Eli Lilly and Genentech, for example, was already on the market, while Biogen's human insulin was still at the research stage; its hepatitis test appeared to be at an earlier stage of development than Centocor's; and, while it had yet to show that its foot-and-mouth vaccine was effective, Genentech's undoubtedly worked.

But, once again, Biogen soon rebounded with what sounded like good news. Several months after it went public, the company reported that it had successfully immunised two chimpanzees against hepatitis-B with its genetically engineered vaccine. Blood tests had confirmed immunity in the vaccinated animals, suggesting that the vaccine would work also in humans.

Hepatitis-B, or serum hepatitis, is a chronic disease transmitted by contact with infected blood, saliva or even semen. In the USA, the total number of new cases may be as high as 150,000 a year, although the incidence of the disease is highest in poor, overcrowded countries. The World Health Organisation reports that there may be as many as 170 million persistent carriers of hepatitis-B in the world. Worryingly, the virus has also been shown to be a cause of liver cancer, one of the more common tumours in humans.

The experimental vaccine used in the Biogen trials was made from hepatitis-B surface antigens produced by genetically engineered yeasts. Three yeast strains had been developed by Murray, working in collaboration with Dr Albert Hinnen, then a research scientist at the Friedrich Miescher Institute of Basel and later at Ciba-Geigy.

Surface antigens are characteristic proteins that lie on the surface of a virus. While disease is transmitted not by these antigens but by the DNA contained in the core of the virus, it is the body's reaction to the antigens that stimulates the production of antibodies and the subsequent development of immunity to the specific virus. So it is unnecessary to inject the whole virus in a weakened or killed state to confer immunity: the surface antigens alone will often work, providing you can produce enough of the right ones.

At the time, hepatitis-B vaccines were made from surface antigens extracted from the blood plasma of human carriers of the disease. The infected plasma is dangerous to handle, however, and all work with it must be carried out in an isolated area by workers who have been immunised against the disease. In addition, each batch of the conventional vaccine must be tested in chimpanzees to ensure that no infectious agents have slipped through the production process.

The dependence of the conventional vaccine on the supply of infected human blood plasma and the rigorous production and testing procedures required have resulted in a costly product which is in limited supply. Total production for one batch of the vaccine takes longer than a year. Worse, products which are made from human blood products have been facing consumer resistance, due to the fear of contamination from other serum-transmitted diseases—notably, acquired immune-deficiency disease (AIDS).

Dr Gilbert estimated in 1983 that it cost over $100 to immunise a patient against hepatitis-B, making widespread immunisation prohibitively expensive and limiting it to such high-risk groups as doctors, dentists, dialysis patients and others likely to come into contact with infected blood. Industry estimates at the time suggested that Biogen's vaccine might cost $10–$30, a significant improvement. Still, Biogen was not alone in the field: Merck was testing a vaccine made from material supplied by Chiron, and reporting that it had already been tried out on 37 volunteers drawn from its own staff.

A $300 Million Threshold

The size of the potential market was a key consideration in Biogen's selection of this particular research target. As Gilbert explained late in 1983, 'we use commercial potential, as well as we can estimate it, as our primary criterion for picking products. In fact, the Scientific Board imposes a restriction that we will

not, to the best of our ability, do any project that leads to a
product that has too small a market. Our criterion is that we, or
our partners around the world, should be able to attain a market
of at least $100 million. Therefore we look for a total world market
of at least $300 million for each product.' Which certainly narrows
down the field.

But how does a company like Biogen proceed once it has
identified a potential product? 'Our products generally go
through thirteen clearly defined steps as they move from an idea
all the way through the research, process development, clinical
trials, regulatory approval stages, and, finally, to the marketplace,' explained Flavell, by now president of Biogen Research
Corporation. The process, he continued,

> begins with the genesis of an idea. We usually start with a
> human protein available in very small amounts. The first step
> is to develop a procedure to recognize the desired products
> through an assay procedure. The assay is used throughout
> research and development to detect and measure the presence
> of the protein. Once we can clearly identify the protein, the
> project moves on to cloning and expression. In this stage, we
> define the necessary genetic material, insert it into a micro-
> organism, grow colonies of the genetically engineered micro-
> organisms and, finally, produce (or express) small amounts of
> the protein to be sure that we have correctly constructed the
> genetic material.
>
> Once cloned, the gene is engineered to express large am-
> ounts of the protein in a microorganism of choice. Once that
> material is being expressed at sufficiently high levels, it moves
> on to process development. The next step is to devise methods
> to extract the material from the microorganism or the medium
> in a pure and safe form that is suitable for use in humans.
>
> When a sufficiently large amount of pure material is avail-
> able, we move on to preclinical testing, where we determine
> in animals whether the material is, in fact, as safe as one
> hopes. Once safety in animals is established, the product
> moves into clinical trials where it is tested on human volun-
> teers. After satisfactory completion of the three-phase human
> trials, it moves on directly to regulatory approval and, finally,
> to the pharmaceutical market.

Clearly, a long-winded process. 'We expect it to take six to eight
years for a reasonably well developed idea to go from laboratory
bench to market,' Gilbert added. 'The initial research and development (the devising of new strains, scaling them up, purifying

the substance, and devising the clinical or commercial process) took us three to four years when we began; it takes us from one to two years now [this was in 1983]. We expect that a few years from now it will take us about a year, but it will still take an appreciable amount of time.'

And that, as we have seen, is far from the end of the story. As Gilbert put it,

> after this initial research is completed, the material has to be placed in clinical trials. Clinical trials take from two to four years. Two years would be extraordinarily rapid; three years is the normal figure and four years or more is also possible. Then it takes anywhere from six months to two years to register a drug with the regulatory authorities after the trials.
>
> The net result is that an idea we start with in the laboratory cannot be a commercial reality for something on the order of six years. For that reason, the products we began working on in 1978 are, at the end of 1983, about to come to market. One of our first products is alpha interferon. Schering-Plough intends to file for registration by the end of this year. That is exactly on our time scale. Any new product that we initiate today will not affect any company earnings until 1990.

This explains the derisive way in which such companies as Biogen and Genentech view many of their competitors. 'We always knew there would be lots of trash companies,' as Gilbert bluntly put it. 'These small companies, with their weak capital bases, have not got the resources either financially or scientifically to survive.'

Biogen itself had grown rapidly. By the end of 1983, it had a total of some 330 employees, over a quarter of whom had either PhDs or MDs. A total of 227 people worked in all forms of research and production, not including the 50 or so people who worked in Biogen board members' laboratories in universities. With its headquarters in a converted watch factory in Geneva and expanding production facilities in Cambridge, Massachusetts, Biogen has also facilities in Zurich and in Ghent. At the time of writing it was building a new facility in Neyrin, near Geneva, to house all the company's Geneva-based activities. Initially this new plant will produce gamma interferon and interleukin-2.

As we have seen (page 59), however, Biogen had been hit by the general cooling of investor interest in biotechnology. It failed to raise the full $40 million it was looking for early in 1984 to fund the commercial development of its gamma interferon and interleukin-2 products: it raised $25 million, and so faced the necessity of investing $35 million, rather than the $20 million originally budgeted, in the planned $60 million R&D partnership. But the reaction would probably have been much worse had it stuck with some of the products it had originally planned to make.

'One of the reasons we develop pharmaceutical products,' said Gilbert, 'is that major products in the chemical industry turn out to have an even longer time frame.' This view represented a significant shift in the company's thinking. As Gilbert summed up,

> When we began in 1978, we thought that recombinant-DNA technology would have broad uses not only in the pharmaceutical industry, but also in the chemical, the food and beverage, and other industries. The discussions we had, and the efforts we made, over the past five years have led all concerned at Biogen to conclude that the most productive use of the current technology is in the pharmaceutical field. We now concentrate our efforts there because we see it as the new field in which the technology will be the most commercially rewarding over the next ten to fifteen years. We do not see the technology being equally effective in the chemical and the food and beverage industries in that period.

This conclusion was largely responsible for Grand Metropolitan's decision to pull out of Biogen: GM is still very much interested in biotechnology, however, and shortly afterwards launched a new company, Biocatalysts (see page 196).

The difference in opinion is more apparent than real, however: while many biotechnologists would agree that it will be some time before recombinant-DNA technology has a significant impact on the food and beverages industry, other novel biotechnologies are already helping reshape the way the industry views its future.

But Biogen considers that its central objective is to make money, preferably lots of it. 'Unlike a contract research company,' says Gilbert, 'our goal is not to make a 10 per cent return on our research effort. Our goal is quite different. We view our research as an investment on which we want to make a ten- or a hundredfold return.'

It sounded impressive, but Biogen had still to make a profit of any kind.

From Gunsights to Genes

To get a slightly different view of what the world looks like from inside the gene factory, take Amersham International—which many consider a biotechnologist's biotechnology company. It happens also to be a company which has been making very considerable profits in the field, despite the fact that some of its core diagnostic technologies are themselves coming under pressure from the emerging biotechnological alternatives.

'We're not a mainstream biotechnology company like Cetus, Genentech or Genex,' said Dr John Maynard, chief executive of Amersham's money-spinning research-products division. 'We provide support products for biotechnology companies. Wherever there is a recombinant-DNA man, we want to give him products to make his job quicker, easier, less skilful. And in that sense our main goal is not to find some new protein which we can sell. We're not looking for the interferons or the growth hormones, but those who *are* looking buy our products.'

But this can be an incredibly difficult field to keep pace with. 'It's very fast moving,' said Dr Brian Ellis, responsible for marketing Amersham's research products, 'and the only way to lead in this area is to have an incredibly fast way of taking information from the market, developing a new product and getting it back into the marketplace.' The fact that Amersham's sales force visits most of the world's biotechnology laboratories, and visits them often (because its products are labelled with short-lived radioactivity), means that its market intelligence is excellent. Unfortunately, its intelligence network has been telling it that significant parts of its product range are doomed.

Most biotechnology companies, however, would love to have Amersham's problems. For a start, the company is the world's leading supplier of research products for biotechnology and biomedical research. When I visited it in 1984 it had just reported that its profits for the previous year had jumped 22 per cent on sales which had themselves risen by 20 per cent. 'Biotechnology —it wouldn't be where it is today without us,' the company was trumpeting at the time in an advertisement in the *Financial Times*. 'Our products helped crack the DNA code which paved the way for genetic engineering.'

In fact, Amersham's roots reach back to the early days of World War II, when the British Government invited a small company, Thorium, to refine radium and manufacture luminous paint for use in compasses, gunsights and aircraft instruments. When peace came, there was a shift in emphasis towards medical applications of radioactivity, based on the manufacture of radium needles, tubes and plaques for use in radiotherapy.

By the mid-1940s, the significance of nuclear energy was clear and new products were being introduced which had been born in the hearts of nuclear reactors and accelerators. This, in turn, led to the formation of the Radiochemical Centre (TRC), a Government-owned body with a mandate to develop and manufacture radioactive products for use in research, medicine and industry. During the 1950s, reactor-produced isotopes became increasingly important, with TRC emerging as a market leader in new research chemicals and pharmaceuticals incorporating these materials.

An isotope is an artificial chemical element produced by bombarding non-radioactive material with neutrons in a nuclear reactor. Isotopes emit characteristic forms of radiation and can therefore be used to track pharmaceuticals and other products around the body. The quantities of product involved are minute, with one run of a cyclotron producing enough isotope to cover one of your small finger nails. Yet this amount is sufficient to 'label' enough chemicals to fulfil hundreds of orders.

The 1960s saw continuing growth, as well as the formation of a US joint venture with G. D. Searle—which venture later became Amersham Corporation. The first clinical assay kit, for insulin, heralded a new era for TRC in the radioimmunoassay (RIA) diagnostic-kit market. Such kits enable clinicians to measure astonishingly small amounts of substances which may be present in body fluids such as blood or saliva.

By the time Amersham, as the company was renamed in 1981, went public in 1982, it had built up a commanding position in the world market for radio-isotopes and radiochemicals. Its 2,000 or so products were being sold to something like 50,000 researchers in 132 countries, about 25,000 of them working in biotechnology laboratories. But the new biotechnology companies, while they have provided Amersham with a rapidly growing market for labelled compounds, restriction enzymes and other tools of the trade, also pose a very considerable—and growing—threat to RIA and other radioactive techniques.

Unlike radioactive products, many of the bio-alternatives have very long shelf-lives, are easy to store, and do not represent a disposal problem; they may also turn out to be very much cheaper. Amersham, in short, realised it had to get in on this new act. It increased its 1983–84 R&D spending to £7.5 million, representing 8.6 per cent of sales income compared with 7.4 per cent the previous year, and announced that the figure would soon be 10 per cent. Its medical products division, whose business is most threatened, was already spending 13 per cent of its sales income on R&D.

Amersham's directors were careful to stress that, by 1990, the company would still be making over half of its income from radioactive products of various kinds. But the writing was on the wall: many of the new generation of diagnostic kits, Amersham was convinced, would be based on an exciting new form of luminescence.

Diseases That Glow in the Dark

If you are designing a test which you hope will throw up a clear warning when a particular virus or other threat to health is present in a patient's body fluids, it is obviously vital that the signal the test sends out is clear and unmistakable. Imagine that you were trying to pick up the virus that causes AIDS. One approach would be to 'flag' with a radio-isotope a set of monoclonal antibodies, each tailor-made to seek out the virus and bind to it. But what if you wanted to adopt a non-radioactive approach? One way would be to use enzymes. Collaborative Research, for example, came up with what it called 'enzyme membrane immunoassay', or EMIA. This approach is based on the use of liposomes, which are synthetic microscopic spheres that can be filled with chemical reagents. Antigens or antibodies are attached to the outside surface of these spheres, which meanwhile have been filled with enzymes. When one of these booby-trapped liposomes meets the appropriate antibody or antigen in a diagnostic system, the resulting reaction punctures the liposome and releases the enzymes. Following the addition of colour reagents, the enzymes can be spotted by use of an instrument called a spectrophotometer.

Other companies, including Cambridge Life Sciences, are likewise working on enzyme-based diagnostic kits. The first launched by CLS was designed to test patients for the pain-killer paracetamol (acetaminophen), often used in suicides. Prompt diagnosis is vital, since the drug can rapidly cause massive liver damage. The enzyme was found during a screening, carried out by the PHLS Centre for Applied Microbiology and Research, of bacteria able to degrade paracetamol. The enzyme finally picked, which came from *Pseudomonas fluorescens*, gives a result in just 10 minutes. Among other CLS targets are tests for antibiotics, aspirin and anti-cancer drugs, to be used to monitor the progress of patients undergoing treatment.

In this approach, the threat is spotted because it triggers a colour change in the test system. Another technique, which companies like Sweden's LKB have been developing, involves 'flagging' antibodies with rare-earth metals that fluoresce when exposed to ultraviolet light. In what is known as 'fluoroimmunoassay', antibodies are tagged with substances like terbium or europium so that they can be tracked and their concentrations determined.

But the approach which has been causing such excitement at Amersham is even more extraordinary. The company put a team of 50 people to work on reviewing the best signal options for use in diagnostic systems. After evaluating well over 20 different techniques, Amersham opted for what is called 'enhanced luminescent immunoassay', or ELIA. This had been developed by Professor T. N. Whitehead and his team at the University of Birmingham's Wolfson Research Laboratories.

'If you look at the various technologies which we can use,' Amersham director Jack Castello explained, 'you've got RIA, which measures both small molecules and large. You've got with basic enzyme technology the capability to measure a portion of that spectrum, from small to large. And with fluorescent technology you've got the capability to measure a much larger portion—but not all. With luminescence, you've got the ability to measure it *all*, plus some.'

The idea of luminescent immunoassay is by no means new. Teams around the world are working on chemiluminescence systems based on the chemical, luciferin, which makes glow-worms glow and fireflies shine. But this approach has suffered from a basic defect: a virus, say, will trigger a pulse of light readily enough, but the pulse tends to be weak and of a duration measured in split seconds. If you want to be sure you see it you have got to invest in some very sophisticated and expensive equipment.

The enhanced luminescence technique, by contrast, rapidly produces a pulse of light which is then sustained for perhaps as long as 30 minutes. In short, the signal is unmistakable. Amersham plans a whole spectrum of diagnostic tests based on this new technology, which will simultaneously cast some much needed light on some dark corners of medicine and, Amersham hopes, put the company one step ahead of the competition again.

FIVE

The New Cures

Slice your finger with a sharp knife and the blood spurts. For most of us such clumsiness is little more than a painful irritation; for some, it can be a question of life or death.

Holding our throbbing fingers under the cold tap, most of us assume that the bleeding will stop in time, as the blood begins to clot. Not so those suffering from haemophilia. This disease, which normally affects only the males of a family but is transmitted by the females, stems from an inability of the blood to form clots. The result is often uncontrolled bleeding, particularly in the joints.

But what causes clotting? The most important clotting agent is 'Factor VIII', which is part of what doctors think of as a 'clotting cascade'. Thirteen main substances, or factors, are involved in this cascade, with different factors coming into play at different stages in the clotting process. Some promote clotting; others prevent the clotting getting out of hand. Without these clotting and counter-clotting systems, as one blood specialist has put it, 'you'd either bleed to death or go solid!'

No one who has tracked the evolution of companies such as Biogen will be surprised to hear that, out of every $100 invested in the equity of US biotechnology companies by the end of 1982, $61 had gone into medical biotechnology, compared with $23 for agricultural biotechnology and just $16 for all other potential applications of biotechnology. To understand why such a high proportion of bio-investment has gone into the health-care sector, you need to look no further than the blood oozing from your own finger. Blood really is miraculous stuff. First, however, a few words about AIDS.

The Blood Bankers

'Three million dollars isn't enough,' said the lawyer representing a 32-year-old US lawyer suffering from AIDS. 'He's a man with a death sentence.' Unusually, however, his client was not a homosexual, like so many AIDS victims; instead, it was alleged, he had contracted the devastating disease from Factor VIII extracted from AIDS-contaminated blood. A haemophiliac needing constant top-ups of Factor VIII, he believed he was another victim of what has become known as 'transfusion AIDS'.

In the ensuing trial it was alleged that the producer of the Factor VIII, Cutter Laboratories, had 'failed to properly screen donors and test the blood for impurities or disease'. Negligence has to be proved in such cases, since in the USA no fewer than 45 states have laws defining blood and blood products as services, not products, a definition designed to shield blood bankers from product-liability legislation. These laws, in fact, were passed during the 1970s, following evidence that another disease, hepatitis, was being transmitted via blood transfusions.

Cutter and other blood-products suppliers were horrified by the implications of such suits. Cutter's own attorney retorted that there was still no certain way to screen for AIDS, arguing that to hold the company responsible for transmitting a new virus 'is a burden the industry can't possibly shoulder'. However, a number of biotechnology companies were already racing ahead with new diagnostic tests designed to pick up the AIDS virus in the blood of donors.

Meanwhile Cutter, a subsidiary of Bayer AG's Miles Laboratories, was exploring other options. Late in 1984, for example, Cutter Biological bought the production and marketing rights to Genentech's recombinant Factor VIII technology. The search for a safer, purer and more plentiful supply of Factor VIII had been long and difficult, but Genentech's unexpectedly quick breakthrough earlier in the year had raised hopes that recombinant Factor VIII would become a commercially attractive alternative to the natural product.

While the existence of this rare protein had been known since the early 1950s, its molecular structure had remained a mystery. Several scientific teams had reported obtaining portions of the Factor VIII gene, but prior to Genentech's announcement, following work on ultra-pure Factor VIII samples supplied by the UK's Speywood Laboratories, no one had reported obtaining the entire gene and making biologically active human Factor VIII. The gene for Factor VIII is very large and the protein product elusive and delicate.

With Genentech's breakthrough, Factor VIII became the largest protein ever to be produced by recombinant-DNA technology —four times larger than human serum albumin, previously the largest recombinant protein and also first expressed by Genentech scientists. Factor VIII turned out to have 2,300 amino acids, compared to 585 for human serum albumin. Most genetically engineered proteins are much smaller: alpha interferon is composed of just 166 amino acids.

Many people still assume that products like alpha interferon and Factor VIII which are now beginning to emerge from the world's gene factories are becoming available for the first time. Not so. Most of them have been extracted from human blood at one time or another, although often in infinitesimally small amounts. For some, indeed, it may remain very much cheaper to extract them from blood than to produce them using genetically engineered animal cells, bacteria or yeasts.

Having visited the Biogens and Genentechs of this world, I decided it was time to take a look at the blood-products industry. I started off with the UK's leading blood-products supplier, the Blood Products Laboratory (BPL). At the time of my visit, BPL was immersed in a £21-million investment programme designed to boost its production of such key products as Factor VIII. Even though companies like Genentech have been achieving near-miracles in the cloning and initial expression of such products, BPL stresses that there are still many hurdles to overcome. Some of the most intractable are likely to be found in the downstream-processing area, the stage at which products are separated and purified (see page 182). Indeed, BPL director Dr Richard Lane warned that

> there are some formidable problems ahead, even though limited expression has taken place for some molecules. After the most difficult bioengineering stage has taken place, which is the full expression of the perfect molecule, it still comes in a solution which none of us would inject into ourselves. It has got to be purified. So it is very much the same as starting with clean plasma, except that you could inject plasma into yourself if you needed to. The onus on purification out of bacteria and yeasts [he stressed] is much greater, and the trade-off is between being injected with the contents of bacterial or yeast hosts, on the one hand, and, on the other, having to put up with some of the current risks from viruses transmitted in plasma.

Clearly, the most worrying of these at the moment is the AIDS virus, but there are others. One of Cetus' first products, for example, was a diagnostic test for cytomegalovirus for use in blood-transfusion centres.

Despite Cutter's interest in Genentech's recombinant Factor

VIII, however, many in the blood industry feel that there may be more traditional answers to some of its problems. The feeling is that contamination problems can be controlled with a three-pronged approach. First, better selection of donors would cut the risk of a blood bank's products becoming contaminated. Second, better plasma-collection methods would help. But, third, assuming viruses still manage to break through, new techniques must be developed to inactivate any that do survive in plasma-derived substances.

Separation and purification are critically important steps in medical biotechnology. Lane was hoping that 'when we are in our new building, the existing building can then be used so that we can gain experience with the downstream purification of fermentation products, even if we just take the product and put it into the fermentation mass and then try to get it back again. We need to find out what fellow-passengers there are and what they do.'

The new genetic-engineering techniques are not going to sweep away the industry which currently extracts proteins from plasma, he predicted.

One could even say that, if you get very clever with your methods for getting products out of bacteria and yeast, those methods are probably going to be suitable for getting products out of plasma anyway. So one shouldn't just discard plasma and say we'll have the new technology from tomorrow. It isn't going to be like that: it's going to be a phased process of development and introduction. BPL's objective will remain the same, to put products in bottles in a form which is safe for patients. It doesn't matter what the starting material is.

The starting material of choice at BPL, it hardly needs pointing out, is blood plasma. The bloodstream of an average adult contains about 10 pints (5.7 litres) of blood. Each drop is made up of 250 million red cells, 400,000 white cells and 15 million platelets, all suspended in the clear, yellowish fluid called plasma. A blood donor in the UK normally gives a pint (about half a litre) of blood and, because of the time the body needs to replace lost red cells, donors usually give blood only twice a year. The plasma, however, is replaced very rapidly, typically in a matter of hours. So the process of 'plasmaphoresis' was developed, which permits the donor to give only plasma. Although this takes longer than normal donations, the re-injection of the separated red cells into the donor's bloodstream means that the frequency of donation can be boosted ten-fold.

BPL receives about 150,000 litres (about 33,000 gallons) of fresh plasma each year, a figure which should rise to 450,000 litres (about 99,000 gallons) when its new production facility comes on stream. Like crude oil, plasma can be broken down into a great many 'fractions'. It has become increasingly common to divide up the basic plasma resource into its various components, so that patients can be treated only with those fractions they actually need.

Plasma costs about £350,000 a tonne (when the blood is given free), so it is vital to extract as much saleable product as possible from the raw material. New separation techniques promise to boost the yield of some products considerably: the efficiency of albumin extraction, for example, is currently about 60 per cent, but the use of certain industrial dyes in affinity chromatography columns (see page 184) can boost that rate to 95 per cent.

Albumin solutions are used to maintain blood-fluid levels in patients with severe burns or with wounds caused by accidents or major surgery. Concentrated albumin is used in the treatment of some liver and kidney diseases, with the UK's National Health Service using about 200,000 bottles a year, extracted from one million blood donations. Genentech is one of the biotechnology companies which has been working on new routes to albumin, in Genentech's case with backing from Mitsubishi. Biogen has been doing work on albumin for a Japanese company, Shionogi.

During 1983, BPL supplied about 40 per cent of UK needs for Factor VIII, with imports running at about 4 million units, valued at £4 million. Imports of albumin ran at about the £3–£4 million mark, with the estimated total bill for imported blood products reaching £10 million in 1983. BPL, in addition, supplies much smaller quantities of Factor IX, another clotting factor.

About 20 proteins are currently extractable from human blood, with the theoretical total thought to be well in excess of 60, many of which have clinical potential. Another key family of blood proteins are the immunoglobulins, the antibodies which fight infections and other diseases. Immunoglobulins are prepared by BPL for use in the prevention of such diseases as hepatitis, measles and mumps. But some antibodies are not present in normal blood donations in sufficient quantities for use in therapy, which means that plasmaphoresis collections have to be made from special donors. These have typically been exposed to a particular infection and have developed the appropriate immune response. This method is used to collect antibodies for use against chickenpox, rabies, rhesus disease and tetanus.

The Protein Telegraph

Three main groups of blood product which are now being worked on by genetic engineers are: *hormones,* including insulin and the growth hormones; *immunoproteins,* including antibodies, antigens (from which vaccines are made), interferons, and lymphokines and cytokines; and *enzymes and other proteins,* including tissue plasminogen activator and collagen. Taken as a spectrum, these proteins promise to help us deal with an enormous number of problems, ranging from cancer to the wrinkled skin of old age.

Traditionally, if you wanted to produce a hormone or some other substance found in the human body, the only way to do so was to extract it from one of the body fluids, such as blood or urine, or to mash up the appropriate organ or gland and try to fish out the particular proteins you needed. The problem with this last approach, illustrated by the extraction of human growth hormone from glands taken from human cadavers, has been that the resulting proteins were scarce and extremely expensive. With any luck, this could all change—and perhaps sooner than we might expect.

Some of the most exciting early applications of genetic engineering have involved hormones, substances which are produced in minute quantities in one part of the body and which, when dispatched elsewhere in the body, can often produce profound effects. We owe the word 'hormone' to Ernest Starling, who in the early years of the twentieth century succeeded in unravelling the secret of these indispensable substances.

It had been clear for some time that the body had some telegraph system which signalled various processes to switch on or off, but it was far from clear what that system comprised. It was known, for example, that the pancreas begins to secrete its digestive juice the moment that the acid food contents of the stomach enter the intestine. Pavlov was convinced that the nerves triggered this response, but Starling, working with Sir William Bayliss, demonstrated that this could not be so: when they severed the nerves leading to the pancreas in animals, the pancreas seemed unaffected.

Then they found that the lining of the small intestine secretes a substance, which they dubbed 'secretin', that triggers the pancreas into action. Starling then suggested the name 'hormone' for all substances which are discharged into the blood by a particular organ to stimulate other organs into action (the name came from Greek words meaning 'rouse to activity'). The discovery of hormones owed much to the work of the UK physiologist Sir Edward Sharpey-Schafer, who had shown in 1894 that an extract of the adrenal glands could raise blood pressure—a finding which led, in 1901, to the isolation by Jokichi Takamine of adrenalin, later recognised as the first pure hormone to have been isolated.

Later still, in 1916, Sharpey-Schafer coined the name 'insulin' for the hormone whose presence he suspected in the pancreatic islets of Langerhans (the word insulin came from the Latin for 'island'), but whose existence was not to be proven for another six years. He was the first to suggest that diabetes might be caused by the lack of such a hormone. The key experimental work was carried out in Canada by Frederick Banting, Charles Best, J. B. Collip and John J. R. Macleod. The first extracts of animal insulin, 'a thick brown muck', were given to diabetic patients early in 1922: they caused fever and other reactions, but also relieved some of the symptoms of diabetes.

Banting and Macleod both won Nobel Prizes in 1923, the first Canadians to do so.

There are currently about one million diabetics in the UK, with an estimated 15 million in the USA—roughly one in every 15 Americans. Of these, 6 million have been diagnosed as diabetics, an estimated 4–5 million are believed to be undiagnosed, and another 5 million are thought to be borderline cases. The outlook is now much better for most diabetics, although the disease is still a blight on the lives of many millions of people—a high proportion of whom suffer from 'late complications', including hardening of the arteries, heart attacks, kidney failure and blindness.

Some of these complications now look like effects of the high-fat diets many diabetics have been persuaded to adopt, but, even if the effects can be reduced by more sensible diets, diabetes remains a major challenge for the medical profession. Eli Lilly, the US pharmaceutical company, has estimated that there are 60 million diabetics in the world, of whom 35 million are in the underdeveloped countries. Three of the approaches to diabetes therapy are synthetic insulin, artificial pancreases, and transplants of foetal pancreas cells.

Traditionally, diabetics have been treated with insulin extracted from pig pancreases. A very small proportion of patients (about one per cent) are allergic to this porcine insulin, so scientists were looking for alternatives. The first came in 1982, when Denmark's Novo Industri worked out how to snip a single amino acid off the end of one of the two chains making up the porcine insulin molecule, replacing it with the human equivalent. This was clearly both simple and effective.

But a more elegant approach was devised by Genentech, working with the City of Hope Medical Center. They genetically engineered bacteria to churn out the two amino chains which then have to be stitched together to achieve human insulin. Sold by Eli Lilly under the trade name Humulin, this genetically engineered insulin has not been a runaway success: doctors have been reluctant to prescribe it, largely because they can see no real advantages over Novo's cheaper modified porcine insulin.

Meanwhile, new approaches are being suggested, including an artificial pancreas, under development at Boston's Joslin Diabetes Center, and grafts of pancreatic tissue from aborted human foetuses, an approach pioneered by Dr Josiah Brown of the University of California at Los Angeles.

The artificial pancreas would still need Novo's or Eli Lilly's insulin, of course, but the transplant approach might mean that some diabetics could dispense with insulin injections altogether. The artificial pancreas, built around a microprocessor and a glucose-sensitive enzyme-assisted electrode (or 'biocensor'—see page 213), would monitor blood glucose and discharge insulin from a rechargable tank implanted just beneath the skin. The insulin would be replenished, by injection through a silicone rubber seal, perhaps at monthly intervals.

The idea with the foetal tissue grafts is that the cells have not yet developed the molecular markers which trigger the rejection response that has been such a problematic barrier in transplant surgery. Clearly, there are ethical issues to be considered here, and there is also the ominous possibility that the pathological process which originally resulted in the destruction of the patient's own insulin-producing cells will begin to destroy the transplanted cells as well. Clearly, Genentech's recombinant insulin faces an uphill battle, but the company has high hopes for its recombinant growth hormone, Protropin.

The Growth Business

'Management had sore butts,' said Genentech's senior product-development director, Michael Ross, recalling the clinical testing of the company's human growth hormone (hGH). As often happens, healthy company executives were injected with the hormone to test its purity and safety. Produced by genetically engineered bacteria, the hormone caused pain and redness when injected into those executive buttocks. Further purification work ironed out this problem before any child was injected with hGH, although some children later developed antibodies to hGH—

forcing Genentech to reformulate the product once more, for fear that this immune response would damp out the growth-inducing effects of the hormone. Again, the reformulation appeared to do the trick.

If your pituitary gland, the pea-sized gland at the base of the brain which triggers growth and sexual maturity, fails to produce enough growth hormone at the appropriate time in your development, you are likely to end up a dwarf, or at least very much shorter than you ought to be. You may also be infertile. Dwarfism has been known throughout history and, until 1958, was effectively incurable. From 1958 on, however, growing numbers of children received hGH extracted from the glands of human cadavers. Unfortunately, a typical two-year course of treatment requires between 50 and 100 pituitaries, which represents a lot of cadavers. Worse, the number of autopsies actually carried out has been falling in the USA, so that the glands have become scarcer. As a result, the US National Hormone and Pituitary Program, which each year supplies hGH free to over 2,000 children out of the 10–15,000 US children estimated to suffer from pituitary dwarfism, has had to ration the drug. Once a boy reaches the height of 5 feet 6 inches (168cm), or a girl 5 feet 4 inches (163cm), the treatment has to stop. Two European companies, Sweden's KabiVitrum and Switzerland's Serono Laboratories, also sell natural hGH, but the price tag of $10–$20,000 for a course tends to put many people off.

Serono, which had become the leading supplier of hGH in the USA and controlled a 50 per cent share of the European market, signed a $1 million hGH deal with the UK's Celltech in 1984. Celltech had developed a version of hGH which is identical to that found in the human pituitary and therefore unlikely to cause the side-effects which have dogged Genentech's product.

KabiVitrum, which has had links with Genentech on hGH through KabiGen, its joint venture with Hilleshoeg and Swedish Sugar, signed up with Biogen for human somatomedin C, which appears to be a critical protein signalling tissue growth following the secretion of hGH from the pituitary. Other substances which are involved in the regulation of growth include the somatomedins (first cloned and expressed in 1982, by Chiron) and growth-hormone releasing factor (isolated, sequenced and synthesized in 1982 by the Salk Institute).

Nearly all of the 80 patents with which Genentech had been issued by 1984 covered methods of production rather than products, and it was having to race to market, in the hope that being first into the fray would count for more than patents. It has also been looking at further applications of hGH, including the treatment of non-pituitary growth problems, such as Turner's syndrome (which affects roughly 1 in every 3,000 live female births); the healing of burns, wounds and bone fractures; and the treatment of osteoporosis, a disease which makes old people's bones even more brittle.

Crashing Reproductive Roadblocks

If you want to produce a particular hormone in bulk, recombinant-DNA technology is far from your only option. In fact, you have four main options.

First, as we have seen, you can extract the hormone you want from human or animal organs, from blood or from urine. Second, some hormones can be made by chemists. The shorter the length of the hormone molecule's amino-acid chains, the more likely it is that it can be produced by chemical synthesis. Such hormones as calcitonin, with 32 amino acids, can be synthesized competitively; somatostatin is another of the roughly 20 small human polypeptides, or proteins, which can now be easily synthesized. Third, hormones may sometimes be produced by animal cells grown in tissue-culture systems. The fourth option is, of course, the one which has been attracting all the publicity, involving microbial fermentation in the wake of genetic engineering.

Among the other hormones which genetic engineers have been pursuing are parathyroid hormone, which could also help in osteoporosis therapy; nerve growth factor, which could help restore damaged nerves following major surgery; and erythropoietin, which helps regulate the development of blood cells. A further rapidly developing area of interest focuses on human reproduction. Targets range from the production of contraceptive vaccines to hormones designed to enable the infertile to have children.

Serono sells a number of hormones used in treating infertility, including follicle-stimulating hormone (FSH), human chorionic gonadotrophin (HCG) and luteinizing hormone (LH). In some cases, however, the problem is not a complete absence of these hormones, but their irregular release. Work at London's Middlesex Hospital and at the National Institute for Medical Research on luteinizing-hormone-releasing hormone (LHRH), which is normally produced by the hypothalamus, has shown how such problems can sometimes be successfully treated—indeed, over 50 women who had been diagnosed as permanently infertile became pregnant, with 19 healthy deliveries and one multiple birth. The hypothalamus secretes a range of hormones which control such biological cycles as waking and sleeping, and the basic problem had been that the gland was not releasing LHRH at the right times to trigger the appropriate stages of the reproductive cycle. The LHRH was pulsed into the women's bloodstreams from drug infusers (originally developed to deliver insulin), the size of tape cassettes, taped to their arms.

Many other hormones are involved in the reproductive cycle: researchers in Montreal and San Francisco, for example, have isolated what they call inhibin from human semen and have shown that it controls the release of FSH, which in turn controls sperm production. The existence of inhibin was first suspected some 50 years ago, when scientists recognised that something was acting as a messenger between the testes and pituitary gland, instructing the pituitary to shut down FSH production when it had produced enough to satisfy current needs. Inhibin may prove to be the basis of a 'male Pill', shutting down sperm production without simultaneously causing impotence. Genentech has been working also on relaxin, which relaxes the birth canal during the delivery of a baby.

Serono extracts its reproductive hormones from body fluids: FSH and LH come from the urine of women who have gone through the menopause, while HCG comes from the urine of pregnant women. The current cost of treating a female patient is about $250–$300 a month, with the current world market for fertility hormones worth perhaps $25–$30 million. Serono expects this market to grow to $100–$150 million by 1990, and has been funding genetic-engineering work at one of the new start-up companies, Integrated Genetics—which believes that it was the first to produce FSH, HCG and LH by recombinant DNA technology.

The Happiness Hormone

Hormone therapy has a lot of complications. Sometimes it may simply be a question of executives with stinging buttocks; other times, as when LHRH was given by nasal spray to both men and women, the result may be hot flushes and a loss of libido —which some may consider too high a price to pay for any contraceptive effect. But other hormones, like endorphin, can be deadly if the dosage is miscalculated.

'Our theory is that, when you jog or exercise or give birth, the euphoria and the analgesia, or "second wind", may occur from the release from the pancreas of the hormone endorphin,' Dr John Houck, president of Endorphin, told *Genetic Engineering*

News. The company, which has a patented process for isolating endorphin, hopes to break into the narcotic drug market with what is sometimes called the 'happiness hormone'.

The name 'endorphin' is a contraction of 'endogenous morphine', reflecting the fact that endorphin seems to perform a natural pain-killing role in the body. Like morphine, first isolated from opium as long ago as 1803, endorphin causes euphoria and analgesia, competing with morphine for the same 'binding sites' in the human brain. Indeed, the most active part of the endorphin molecule, a sequence of five amino acids called enkephalon (from the Greek for 'in the head'), binds even more tenaciously than morphine does.

Relatively little is known, even today, about the way such peptide drugs work in the body. All peptides are fairly simple molecules, short polymers from 3 to 60 amino acids in length, each having a different geometric shape. As each peptide is released by the relevant organ, often in response to outside stimuli, it begins to look for appropriately shaped receptors in the brain or other tissue. Once it has found a suitable receptor, it typically fits in snugly, like a piece of a jigsaw. Once in place, it may transmit some form of signal, which then promotes the release of other substances.

Enkephalon is part of beta-endorphin, one of the three hormones which go to make up the full endorphin molecule, composed of a total of 31 amino acids. An early problem with beta-endorphin was that it failed to do very much, partly because it seemed to find it very difficult to cross the barrier between the bloodstream and the brain, and partly because, once in the brain, it was rapidly deactivated by the brain's own enzymes.

One man who thought he might have the answer was Endorphin vice-president Dr Charles Kimball. He believed that endorphin would be found in the placenta: it was—and it proved to be a different type. This molecule, Houck recalled, proved to be 'four times bigger than the pituitary molecule. Most important, if you inject it intravenously, it works just fine. It breaks through the blood-brain barrier, gets into the brain and survives. This is possible,' he suggested, 'because it is a big compact molecule that is resistant to the various enzymes that would normally consume it.'

But which of the four options should be used to mass-produce the active parts of the endorphin molecule? Beta-endorphin, Houck concluded, was 'too big to build by synthetic methods of hooking one amino acid to the next'. Instead, Endorphin has been working on recombinant-DNA approaches in an attempt to produce enough endorphin for clinical trials.

Interestingly, doctors suspect that many joggers actually become addicted to running because, as they burst through the 'pain barrier', their pancreas releases endorphin into their blood, producing the 'runner's high'. When one of the world's leading advocates of jogging, Jim Fixx, died of a heart attack on his daily run, the consensus was that he had become addicted to exercise, out-running his body's ability to run. Doctors also know, however, that if they can get a depressed patient to go jogging, he or she can often show a marked improvement. This fact suggests that endorphin may play an important role in the treatment of depression and other psychological ailments.

Houck suspects also that acupuncture, which involves inserting needles into patients, may work, at least in part, by triggering the production of endorphins. Take a patient on whom acupuncture is having the desired analgesic effect and inject a competing peptide drug, he has pointed out, and acupuncture 'doesn't work any more'.

One advantage that endorphin may offer over morphine in medical uses is that it does not appear to have such a strong effect on a patient's breathing. In mice, for example, the amount of morphine needed to produce analgesia is fairly close to the amount needed to make the animal stop breathing. Even so, endorphin would need to be carefully prescribed. Given to test animals in larger amounts, it can kill in seconds.

Such pain-killers are badly needed, however, in the treatment of cancer patients and of others suffering tremendous pain. Biotechnology could help blunt cancer's dreadful edge in other ways, too. The interferons, which were among biotechnology's first 'super-star' products, are just some of the body's own defence mechanisms which are being mobilised and reinforced in the battle against cancer. Unlike antibiotics, whose medical use is rather like winning battles against infections by drafting in whole battalions of mercenary soldiers, some of the lymphokines and cytokines now being developed work by boosting the ability of the body's own immune response to do the job.

$50 Million an Ounce

'Why don't we just pull out all these tubes and let me go home?' was the heart-rending question a 12-year-old American boy asked shortly before he died. 'David', who spent all but 15 days of his life in plastic, germ-free bubble tents, was unique: no one else had survived so long without a functional immune system. His very survival resulted from the fact that the family doctor knew that an earlier brother had died of the same problem, and ensured that the newborn child was promptly insulated from all sources of infection.

His only real hope of a long-term cure lay in a transplant of bone marrow from a genetically matched donor. Unfortunately, David's older sister's marrow cells did not match his, but they were nonetheless used, after treatment, in a transplant operation. Early in 1984, the boy crawled through his air lock and into his mother's arms. Two weeks later he was dead.

His fate made world headlines, in a way that the deaths of uncounted millions of Third World children have failed to do. David actually died from the explosive, cancer-like growth of his white blood cells, known as B-lymphocytes, a condition which has affected other transplant patients. His doctors stressed, however, that such children are perhaps 10,000 times more likely to develop such blood cancers.

Here was a case, however, which underscored the infernal complexity of immune-system diseases. When interferon was first mooted as a possible cure for cancer, its value was a staggering $50 million an *ounce* ($1.8 million per gram). In those early days, you would have needed to process thousands of tonnes of blood to extract an ounce (28g) of interferon. Yet a flourishing black market soon developed, as wealthy cancer victims tried to track down a cure. As film actor John Wayne lay dying of cancer, his fans talked of pooling their resources to buy him interferon; and sensitive—and ultimately unsuccessful—diplomatic negotiations went on to obtain enough of the substance to treat the dying Shah of Iran.

The interferons are believed to be the body's first line of defence against viral attack. Unlike bacteria, which can be destroyed by antibiotics such as penicillin, viruses have defied our best attempts to find a cure. The key to their success is that they are primitive genetic engineers, invading and taking over our cells. Breaking through the cell wall, a virus inserts its own programme, in the form of nucleic acid, into the cell's machinery. So, instead of turning out the substances the cell needs to function, this machinery starts churning out more viruses. When the invaded cell bursts, these viral particles are released into the bloodstream, invade new cells—and the cycle repeats itself.

Trying to destroy viruses inside such cells has been impossible without destroying the cells themselves. Our immune system does have a number of lines of defence against viral attack, however. Certain white blood cells, for example, can learn to produce antibodies which bind to virus particles and neutralise them. Others can learn to recognise virus-infected cells, by detecting 'surface proteins' embedded in the membrane of the infected cell, and kill them directly. But these two processes can take up to two weeks to become effective. The immune system is slow to react, although it can often prevent a further infection. Vaccines, as we shall see, 'educate' the immune system to recognise and attack a specific virus. The weakness of vaccination is that each virus tends to need a different vaccine, yet, for example, up to 100 different viruses are thought to be able to cause the common cold. Other viruses, such as the influenza virus, can mutate in such a way as to slip past the antibodies built up by previous bouts of 'flu. Indeed, it is chastening to recall that, while vaccines may help us *prevent* such viral diseases as chickenpox, measles, polio or smallpox, there is still no known way of *curing* such infections once contracted. All you can do is try and help the patient last out the disease.

The interferons, by contrast, seemed to offer cures for viral diseases. The discovery of this family of substances dates back to 1956, when the UK virologist Dr Alick Isaacs, working with a Swiss colleague, Dr Jean Lindenmann, tried to find out why human subjects, and isolated cells in test-tubes, could be infected by only one virus at a time. They injected chick-embryo cells with influenza virus and noticed that the infected cells were releasing a protein into the culture medium; this protein enabled uninfected cells to resist viral infection. In their report, published in 1957, Isaacs and Lindenmann called the substance 'interferon', because of this apparent ability to interfere with the way a virus propagates itself. They also pointed out that interferon seemed to be effective against a *range* of viruses, which really was something new.

For many years little further progress was made, largely because interferon was very difficult to extract from human blood —indeed, many people doubted that the substance existed. In 1962, however, it was found that interferon inhibited the growth of certain tumour cells cultured in a test-tube, and further research showed in 1970 that interferon could actually shrink tumours in experimental mice. This potential as a cancer treatment attracted many other teams into the race.

The first real progress came when Dr Kari Cantell of the Finnish Red Cross found a way to produce enough interferon for clinical tests during the 1970s. But, even so, extraction was still very difficult and expensive. The process he developed made interferon by taking white blood cells (leukocytes) from blood donors, infecting them with a virus, and collecting and purifying the interferon produced as a result. This technique could squeeze only half a gram of partially purified interferon from more than

50,000 litres (about an ounce per 630,000 gallons) of blood plasma.

In fact, the protein's structure and properties remained a mystery until 1980, when a team led by Dr Charles Weissman, a member of Biogen's board, announced the isolation of the interferon gene and promptly began synthesizing interferon. 'In a few months,' Biogen recalls, 'more was learned about interferon biochemistry than had been discovered in the previous 23 years.'

The immune system's astonishing sophistication was beginning to be discovered. For example, while it can block the growth of some tumour cells, interferon does not greatly affect the growth of most other cells in the body. The interferon secreted by the first wave of cells to be infected binds with other cells, making them virus-resistant. And it also stimulates special white blood cells, 'natural killer' cells, which seek out and destroy tumour and virus-infected cells.

More discoveries will obviously follow as further elements of the immune system are identified and unravelled. There has been much interest, for example, in the lymphocytes, which come in two types: B-cells and T-cells. Among recent breakthroughs, scientists at the universities of Stanford and Toronto have unravelled the genetic sequence coding for the part of T-cells that 'recognises' invading antigens—which may be those protruding proteins on the surface of a virus-infected cell.

These T-cell genes turn out to be closely related to the genes used by B-cells, which are much better understood. The B-cell part of the immune system is based on lymphocytes, which carry antibodies on their surface. Confronted by a viral invader, these B-cells multiply rapidly, pouring out antibodies able to destroy the foreign protein, or antigen, together with the virus or other threat to which that antigen is attached. T-cells are more complicated: they deploy a family of helper, suppressor and killer cells, in addition to helping B-cells to perform *their* tasks.

Victims of AIDS have been found to have a very low count of these T-cell 'helpers', which opens up the way for infection. AIDS often kills by letting in diseases which an unimpaired immune system could deal with easily. It also seems that the reason that the drugs given to transplant patients have the same effect is that they, too, damage the T-cell system.

And it is here that the promise of the interferons and other substances extracted from the body is still seen as considerable. Unlike the drugs and radiation used to treat many cancers, including those which plague AIDS sufferers, interferon does not kill normal cells. At last there is the chance of aiming a drug specifically at cancer cells, leaving the rest of the body unharmed.

Three classes of interferon have been identified: alpha (or leukocyte) interferon; beta (fibroblast) interferon; and gamma (immune) interferon. There are more than a dozen types of alpha interferon alone, all produced by leukocytes and all moving easily from cell to cell in the body. Beta interferon, which is produced by a number of cell types, including the fibroblast cells which make up connective tissue, does not move freely in the body; it tends to remain where it is injected. Gamma interferon is produced by white blood cells in the spleen and in the blood-stream during the immune response.

Not surprisingly, these different forms of interferon are proving to have different effects. Each form must be separately tested against a range of diseases, and then retested in various combinations to look for improved effects. In the three years from 1980, the US government and industry pumped something like half a billion dollars into interferon research. Some analysts suggest that a similar amount will be needed to pin down the potential applications of interferon most likely to be worth pursuing. One described it, somewhat unfairly, as 'a drug in search of a disease'.

Moulding a Magic Bullet

While companies like Hoffmann-La Roche and Schering-Plough have poured millions of dollars into interferon research, others have fought shy. Over 50 companies are active in the field, but their R&D expenditure is typically very much less than these two great companies have been spending. And even Hoffmann-La Roche and Schering-Plough recognise that a good deal of the original glamour has rubbed off interferon. 'I don't think anyone can now say, as they did a few years ago, that interferon is a magic bullet—a drug that works against all cancers all the time,' commented the head of Hoffmann-La Roche's interferon testing programme, Dr Zofia Dziewanowska. 'But it will have a place. That is certain.'

The two companies expect a significant return on their investment even if interferon, to quote Schering-Plough president Robert Luciano, 'goes down the tubes'. As a senior Hoffmann-La Roche vice-president told the *Wall Street Journal*, 'our interferon programme has been a catalyst. It has facilitated most of what we know about identifying such molecules and producing large amounts that we can study. Down the road, many of the new drugs will be developed the way we're developing interferon.'

The first major patent battle was over alpha interferon (page 39), but gamma interferon looks much more powerful than either alpha or beta. By the time Biogen got its European patent on alpha interferon, its range of targets had narrowed considerably. Biotechnologists no longer saw alpha interferon as a potential cure for such common cancers as those affecting the breast, colon or lung, but early clinical trials suggested that it was at least as effective as other therapies against cancers of the blood, bone, kidney and skin. It also looked promising as a treatment for various forms of herpes, for multiple sclerosis and perhaps even for AIDS.

The potential scale of the market for successful products in some of these areas could be very considerable. Take the herpes viruses, which often hibernate in nerve roots, lying dormant until another outbreak is triggered. There are many forms of herpes virus, although the one which has attracted most publicity has been *Herpes genitalis*. It has been estimated that 20 million Americans have had this infection, and that somewhere between 300,000 and 500,000 more contract it each year. Although it is not fatal in adults, *H. genitalis* can cause mental retardation or fatal brain infections in babies exposed to it at birth. The early indications were that interferon-based ointments might be useful in controlling some herpes infections.

Biogen's alpha-interferon-testing strategy has involved aiming for groups of patients who, in effect, have nothing to lose by trying the drug. These include patients suffering from cancers for which there was no existing treatment, and, used as a cold preventative, high-risk groups of patients, such as the aged, asthmatics and those suffering from bronchitis or emphysema, for whom a cold can be a life-threatening event.

Interferon is one of a family of molecules, the lymphokines, produced by white blood cells and involved in all aspects of the immune response: altogether, there are thought to be at least 30 and possibly over 100 lymphokines. Others, aside from interferon, include colony stimulating factor, the tumour necrosis factors and macrophages activating factor. Cytokines, which seem to have effects similar to lymphokines, include the thymic hormones, associated with the thymus gland. The world's genetic engineers have been rifling through some of these other lymphokines and cytokines in search of potential drugs. A great deal of research is still needed in this area, however, because the action of cytokines is so complex. One cytokine, for example, may cause a cell to change and make a second cytokine, which then turns on production of a third cytokine in another cell. Even more confusingly, the same cytokine can have different effects on different cells.

The most important of the lymphokines to date has been interleukin-2 (IL-2), first described in 1976 by Dr Robert Gallo and his team at the National Institutes of Health, Bethesda, Maryland. 'We knew it was important,' Gallo later recalled. 'We just didn't know how important.' It seems to be a promoter of T-cell activity, which has excited interest in its use in AIDS therapy. IL-2 is another messenger, or signal, molecule used by white blood cells to regulate various aspects of the immune response. It also, apart from promoting T-cells, activates natural killer cells. As with the interferons, many of the new biotechnology companies are chasing IL-2.

Genentech, meanwhile, had come up with what looked a very promising anti-cancer agent, lymphotoxin, which can selectively destroy malignant tissue. Unlike any other such agent found to date, lymphotoxin, which is one of the tumour necrosis factors, can break apart the membranes of tumour cells—and may also have a secondary role in stimulating the body's defences against cancer. Lymphotoxin appears to boost the effectiveness of gamma interferon, which calls the body's macrophages into action to gobble up cancer cells.

Although it had seemed that Cetus and Hoffmann-La Roche, which had been collaborating with Immunex, were well ahead in the IL-2 stakes, Genentech retorted that lymphotoxin could well dispense with the need to prescribe IL-2 at all. The company believed that lymphotoxin acts by stimulating white blood cells to produce IL-2.

Among the other products currently under development is Protein A, which boosts the immune system and is a natural component of cell walls. Sweden's Pharmacia extracts Protein A, worth perhaps £10,000 a gram (nearly £300,000 per ounce) in 1984, from the cell walls of the bacterium Staphylococcus aureus, using a batch method of production. The UK's Fermentech, by contrast, has had its sights on a continuous fermentation process. It is thought that Protein A could have important applications in the treatment of some immune diseases where patients' own antibodies attack some part of their bodies—as in rheumatoid arthritis. Protein A could help remove some of the offending antibodies.

As far as 'magic bullets' go, however, what are known as immunotoxins represent one of the most promising approaches. Antibodies, as we have seen, are an important part of the immune response. Monoclonal antibodies (see page 28) are highly specific in the targets they seek when injected into the bloodstream, a characteristic which has suggested an ingenious new approach to the treatment of cancer and other crippling diseases caused by rogue cells.

Antibiotics can be considered a crude form of 'magic bullet', in that they leave normal body cells largely unaffected, damaging only bacterial cells, whose metabolic processes they disrupt. But a more sophisticated approach, which is being explored by such companies as Cetus, exploits the ability of monoclonal antibodies to home in on just one group of targets. The basic idea is simple: if you can come up with an antibody tailor-made to bind with the type of cell you want to destroy, you can hook a highly toxic molecule, like ricin, to it. Like a computer-guided torpedo, the antibody carries the deadly cargo straight to the rogue cell.

'It's essentially a guided warhead,' said Cetus president Robert Fildes. 'The monoclonal antibody finds the target you want.' Cetus is confident from laboratory experiments that the approach will work. 'You can take an immunotoxin made up of a monoclonal antibody specific for breast cancer coupled to a toxin molecule, put the immunotoxin in the presence of the breast cancer cells and it will kill at extremely low concentrations.' Cetus scientists have screened over 20,000 monoclonals to pin down the ones which are specific to such important cancers as those affecting the breast, colon, lung and prostate gland.

But monoclonals, like the interferons, have not proved quite as useful as pioneers in the field expected. They have had their undoubted successes, helping to define the structure of oncogenes, the rogue genetic material now thought to trigger many cancers; they have been used to control tissue rejection in transplant patients; and they are at the heart of many of the new (and much more accurate) diagnostic kits which are beginning to pour out of leading firms like Hybritech, the first company founded for the sole purpose of producing and selling monoclonals. Yet there are flies in the ointment. 'People have underestimated the considerable problems in commercializing monoclonal antibodies,' explained Nigel Webb, president of Damon Biotech. Part of the problem is that many of the target cells have proved to be unexpectedly evasive. 'Some cancer cells,' said Frank Rauscher of the American Cancer Society, 'have the diabolical ability to pull in their antigens and protect themselves.' Worse, tumour cells of the same type may differ from patient to patient, and even the most carefully tailored monoclonal can get confused.

Monoclonals, nonetheless, will prove a tremendous boon both in research and in the treatment of a wide range of diseases. But as far as the sheer numbers of people likely to be protected are concerned, an even more significant approach is vaccination. This basically involves educating the body's own antibodies to recognise a particular intruder—and to mobilise B-cells, T-cells, macrophages and the other elements of the immune response in the body's defence.

The Super-Vaccines

If Edward Jenner were still alive and attempted to repeat the experiment which made him famous, he would almost certainly be imprisoned. The experiment, during which he inoculated 8-year-old James Phipps with pus from a milkmaid's cowpox pustule and then, two months later, with a potentially fatal dose of smallpox, violated even the laxest of today's medical research guidelines. But the boy lived and, thereby, helped revolutionise medicine.

Genetic engineering promises to effect a further revolution in vaccine technology, greatly accelerating the speed with which new vaccines appear and dramatically cutting the risks involved in immunisation.

In retrospect, however, it emerged that not only was Jenner decidedly lucky that his first patient did not enter history as his victim, but he also happened to pick a disease with some fairly uncommon characteristics, which made smallpox a comparatively easy first target. To understand why the recombinant-DNA approach to vaccines represents such a radical departure, it is worth briefly recalling one or two of the developments in immunology since 1796, the year when Jenner took James Phipps' life in his hands.

As so often with scientific revolutions, Jenner's did not take place in a vacuum. The Chinese and the Turks had long known that, if you contracted smallpox from someone with a mild case, your chances were better than if you got it from someone with a virulent dose. In fact, inoculation had been tried before Jenner came on the scene, using pathogens taken from smallpox victims, although this approach was rather like playing Russian roulette with several of the pistol's chambers loaded.

Jenner, by contrast, had observed that milkmaids infected with cowpox acquired an immunity not only to cowpox but also to smallpox. A similar immunity could be acquired by grooms and stable-lads to a disease of horses called 'the grease'. But even when he had demonstrated the practical advantages of his technique, Jenner had precious little idea how it actually worked.

The jigsaw only really began to be put together when Pasteur, born the year before Jenner died, came up with his germ theory of disease. Once it was recognised that exposure to a specific pathogen could cause a specific infectious disease, the body's ability to learn how to cope with that pathogen through earlier exposure to similar but weaker (or dead) pathogens began to make sense.

Three main types of vaccine are now used: toxoids, killed vaccines, and attenuated living vaccines. Toxoids are extracts of toxins excreted by pathogenic bacteria, such as those causing diphtheria. The toxins are neutralised with formalin and stimulate the formation of antibodies when injected into the body. Cholera, polio and typhoid are examples of diseases controlled by killed vaccines, generally prepared from bacteria or viruses which have been chemically treated. Attenuated live vaccines, by contrast, are prepared from cultures of micro-organisms that have become less virulent, sometimes because they have been grown in an unusual host, such as a horse.

In each case, the immune system is responding to a foreign protein, or antigen, by producing antibodies. Once produced, the antibody-producing systems tend to stay around in the bloodstream, triggering a much faster response next time around. But what if the pathogen, instead of being stable as in the case of measles or smallpox, is unstable and readily changes its coating of surface proteins? Clearly, it emerges with a different set of antigens and can hoodwink our antibodies. Influenza is the most obvious illness caused by such turncoats.

Viewed from the perspective of 1796, of course, immunisation has brought astonishing benefits and new vaccines are still coming through—like the chickenpox vaccine developed by Merck, based on a strain of live virus first isolated in Japan in 1974. Some see this vaccine, which tackles the chickenpox herpes virus, as a natural first step to a genital herpes vaccine.

But, as the controversy over the whooping-cough immunisation programme has shown, the use of conventional vaccines can introduce a totally new set of problems. These begin to loom large once the vaccine has brought a particular disease under control, when we face the trade-off between a falling number of deaths from the disease and a now relatively high number of deaths or serious illnesses caused by the vaccine itself. And attenuated vaccines can entrain an even worse problem: if the attenuated organism reverts to its virulent form, the vaccine may actually cause an epidemic of the disease it was designed to prevent. Recent outbreaks of foot-and-mouth (hoof-and-mouth) disease, for example, have been traced back to contaminated vaccines used in continental Europe, where vaccination is preferred to the UK approach of slaughtering infected herds.

Another main area of concern, which is tipping the scales in favour of the genetic engineers, centres on the fact that some pathogens used in immunisation programmes are grown in monkey cells in culture. There is a possibility that they could accumulate tumour viruses while in culture and then transmit them to human patients.

It is easy to understand the genetic engineer's interest when you recall that antigens are often proteins, produced by the pathogen, and that genetic engineering involves manufacturing proteins to order. If the foot-and-mouth immunisation programme had used bio-engineered vaccines, there would have been no vaccine-induced epidemic because such 'sub-unit' vaccines would simply have consisted of tailor-made antigens, rather than of a virus which had been inadequately treated.

Some of the advantages of such genetically engineered vaccines are discussed on pages 66–7, in relation to Biogen's hepatitis-B vaccine. Many other companies are now in the race. 'A great deal is now known about how to identify surface antigens,' Cetus' Robert Fildes explained. 'Once an antigen is identified, then the genetic engineering required to reproduce it is relatively straightforward.' The earliest sub-unit vaccines on the market, however, have been designed for animals, like Cetus' own vaccine designed to prevent scours in piglets (see page 133). In the USA, it takes just two years to get such new veterinary products approved by Federal regulators, compared with five to seven years for a human vaccine.

Given the scale of the expense likely to be involved in bringing such vaccines to market, commercial biotechnologists have been understandably cautious in the targets they have picked. They have tended to go for Western diseases, like genital herpes, rather than for more serious diseases which, like malaria, affect much larger numbers of much poorer people. Despite hopes that malaria was on the verge of eradication, following the invention of DDT and chloroquine, a synthetic version of quinine, many strains of malaria-carrying mosquito have developed resistance to DDT, while *Plasmodium falciparum*, the most lethal of the four parasites which cause malaria, developed resistance to chloroquine. Today, more than half the world's people live under the shadow of malaria, with an estimated 250 million falling ill each year and a million dying every year in Africa alone. 'We have moved from despondency to euphoria and back again,' said the director of the World Health Organization's tropical-disease research, Dr Adetokunbo Lucas.

A key problem with malaria is that each of the three main stages of the parasitic cycle is marked by a different antigen. The *Plasmodium* parasite transmitted by mosquitoes is effectively three bugs in one: the sporozoite enters the blood when the mosquito first penetrates the skin; the merozoite invades red blood cells and causes the fevers and chills which bedevil malaria victims; and the gametocyte which, ingested by the next wave of mosquitoes, triggers the repeat of the whole cycle. Antibodies developed against the antigens characteristic of one of these stages are obviously not going to work against the others.

Some scientists continue to question whether a malaria vaccine is even possible, since the sporozoites injected by a mosquito take only a few short minutes to find shelter, in the victim's liver, from the prowling antibodies in the bloodstream. This does not give the antibodies much time to react, and malaria will result if even a few sporozoites break through to the liver. The consensus among malaria researchers is that the best approach would be to use a vaccine consisting of a 'cocktail' of all three major antigens.

Once, the only way to find the sporozoites from which you needed to extract such antigens was to dissect the salivary glands of infected mosquitoes, but genetic engineering is beginning to make real headway. Among recent developments was the announcement by groups of scientists at the New York Medical Center, the National Institutes of Health and the Walter Reed Army Institute of Research that they had synthesized the antigen which can trigger immunity to the sporozoite stage of the disease. Merozoite and gametocyte vaccines are also under development.

An experimental vaccine against all three stages could be available in the 1990s. 'If it works,' said one research scientist, 'malaria eventually will be eradicated like smallpox.' Such vaccines will play a major role in raising the standard of living in the developing world, if the political will can be mobilised.

'How long,' asked one doctor in Mali, 'can the international community tolerate a situation where men go to the moon with ease—but find it too difficult to organise a vaccination campaign against measles in the Sahel because it needs thermoses and fridges?'

Prescribing the Future

Unquestionably, however, a revolution is taking place in medicine. It is a revolution which will affect the lives of most of us. 'Biotechnology in itself is no longer what's exciting,' Robert

Fildes remarked. 'What's exciting is the products we've actually begun to produce.'

The executives of the world's genetic-engineering companies will be injecting themselves with a growing number of new products in the next few years. Schering-Plough president Robert Luciano was one of those who volunteered himself as a guinea-pig, using an interferon nose spray twice a day for a month. Unlike many others who took part in the trials, some of whom thought that the side-effects were worse than the cold they had been trying to avoid (see page 65), he experienced no side-effects —and did not catch a cold. Although he admitted that his inclusion in the test 'doesn't mean a hell of a lot', he stressed that 'if just 10 per cent of cold-sufferers can use interferon, we've got a heck of a product'.

Nasal inhalation, meanwhile, is one of the new drug-delivery systems being developed. Nasal sprays, says California Biotechnology, which is working on one system with Boston's Beth Israel Hospital, 'could be particularly important to the commercial development of many products generated by biotechnology since none can be administered in pill form. Almost all of the genetically engineered products are protein molecules which, if taken orally, are destroyed in the stomach before entering the bloodstream. Like insulin, these drugs must now be administered by injections.' Hands up those who would prefer to be treated with a spray.

Despite the inevitable disappointments with many of the much-hyped 'magic bullets', doctors are now beginning to work with the first of a wave of new products. Among the fastest onto the market, because they do not have to go through such rigorous testing as products which are actually taken into the human body, are the growing variety of diagnostic kits. Many of these are based on monoclonals and cut the time involved in testing patients for particular diseases from the days and weeks once needed to culture up the offending bacteria, to just hours.

As far as infectious diseases are concerned, the first diagnostic kit on the market was the test for chlamydia, which has become the leading sexually transmitted disease in the USA. This kit was produced by Genetic Systems and is marketed by Syva, a Syntex subsidiary: it is selling well at about $10 apiece. Inevitably, the price for such kits will fall as the market matures.

Another group of tests are based on the use of DNA probes (see page 22). While some analysts have been trying to predict which of the many techniques now being developed will ultimately win the market, the chances are that many will find a niche of some sort, for particular applications. 'To say that either

DNA probes or antibodies will win out is crazy,' said Howard Greene, chairman of Hybritech, which has established an affiliate company, GenProbe, to exploit DNA diagnostics. 'That's like saying, "Who's going to have it, MacDonald's or Colonel Sanders?"'

One area where DNA probes do appear to have something rather extraordinary to offer, however, is in the diagnosis of genetic diseases. Many such diseases are so dreadful that parents, given the option, would choose to have a foetus aborted if tests showed it would be a victim. The problem is that only a few laboratories can carry out the sophisticated tests needed, and these often take weeks to complete—by which time an abortion may be out of the question. Now companies like Cetus are beginning to launch tests which cut the waiting time to just a day or two. Cetus, for example, has been working on a test for sickle-cell anaemia.

And this is where some of the doctor's (and parent's) trickiest ethical problems are going to be encountered in future. As we shall see in Chapter 12, these new techniques will provide us with new information which some of us would rather not have. And what will employers do when they have information suggesting that particular employees, some destined for top jobs, are genetically susceptible to heart attacks?

For those who have escaped such scrutiny, but suspect that a heart attack lies just over the horizon, there is a certain amount of cheering news. One of the proteins which companies like Genentech have been working on could be used to treat thrombosis and heart attacks, the leading causes of death in the industrialised nations. Another natural protein found in minute quantities in the blood, tissue plasminogen activator is a potent, highly specific substance which can dissolve blood clots in the coronary arteries. It is one of a family of 'fibrinolytic' enzymes, a number of which may prove to have therapeutic applications.

Alternative approaches involve the use of two enzymes, streptokinase (extracted from *Streptomyces* bacteria) or urokinase (extracted from urine or cultured human kidney cells). The problem with streptokinase, however, is that it can produce an allergic reaction, while both streptokinase and urokinase, known as 'thrombolytic' enzymes, activate the clot-dissolving system throughout the body, so that internal bleeding may result.

Other products which may eventually appear in the doctor's surgery are genetically engineered antibiotics, with some scientists also now looking for antibiotics which not only kill invading germs but also boost the body's own immune system. By helping doctors cure such diseases as pneumonia, scarlet fever, syphilis and tuberculosis, early antibiotics transformed the face of medicine, but some of their peripheral effects tended to be ignored in the rush. It has been known for years, for example, that certain antibiotics change the shape of some bacteria, and there is at least a possibility that such antibiotics could be used to make bacteria more vulnerable to the body's own natural killer cells, such as phagocytes. Moreover, some of these cells seem able to absorb antibiotics, thereby enhancing their own effectiveness against bacteria.

Developing new antibiotics used to be rather like finding a heavier hammer to crush a beetle: you found one which worked and mass-produced it. Increasingly, however, the medical profession recognises that there can be no total victory over pathogenic bacteria. 'We now know that the best we can do is stay even with these creatures,' said Richard White, director of antibiotic research at Lederle Laboratories. 'To do this, we will need a better understanding of how antibiotics work and of the mechanisms by which bacteria develop resistance to them.' The new genetic-engineering techniques are already helping identify bacterial Achilles' heels, knowledge of which will permit a much more sophisticated approach to the treatment of infections.

Phage therapy, which exploits the vulnerability of many infectious bacteria to even smaller microbes, or phages, will also help this drive towards greater sophistication. Many chronic infections refuse to respond to antibiotics, but research with phages has shown that these highly specific killers can be used to treat some very unpleasant infections.

A Cure for Ageing?

'I hope I die,' one anthem of the 'sixties ran, 'before I get old.' Short of Armageddon, however, the chances are that more of those alive today will survive to a ripe old age than at any previous time in history—a trend with staggering social and economic implications. Indeed, anyone who thinks that the biotechnology industry will soon run out of ideas should take a look at such new companies as Senetek. A world of senior citizens, they are betting, will be a world crying out to buy products which slow, or at least mask, the ageing process. Senetek's share issue in 1983 triggered considerable interest among investors, who, when told that the company planned to tackle the ageing process, or what it termed 'senescence', applied for more shares than were actually on offer.

No one is yet promising elixirs of eternal youth, but the gradual banishing of one disease after another is shifting the spotlight to afflictions which are closely tied to the ageing process itself. If disease can be held at bay by some of the bio-medicines we have already looked at, the natural human lifespan appears to be of the order of 90 to 100 years and, with improvements in living conditions and advances in medical science, an increasing number of people are hitting this ceiling. Ultimately, perhaps, people may come to enjoy a life expectancy even longer than this, but meanwhile biotechnologists are thinking about ways in which the apparently inexorable ageing process might be slowed down.

Among the diseases which are believed to be associated with ageing are diabetes, mental disorder, deficiencies in enzyme and hormone production, and defective genetic repair mechanisms. And there is growing evidence that many of the debilitating problems which come with advancing age are under genetic control—with relatively few genes involved.

One important clue turned up recently when researchers in the USA and Denmark noticed that the activity of a particular protein, elongation factor 1 alpha, changes as a cell ages. This factor is found in fairly large amounts in the protein-producing cells of all mammals, catalysing a key step in protein synthesis.

If you isolate a human embryo cell and grow it in a culture medium, you find that cells from tissues such as the skin have a finite lifetime. When a group of normal human embryo cells reaches the end of their ability to proliferate, it has been found, they do not simply stop dividing. Instead, the time necessary for a doubling increases as the population of cells approaches a limit of 50 doublings.

Professor Brian Clark, based at the University of Aarhus, Denmark, leads a research group which recently spotted a key link in the ageing of cells. As a cell ages, the level of enzyme activity inside it falls and the ability of elongation factor 1 alpha to catalyse protein synthesis begins to go awry. Clark, a director of Senetek, has refined two analytical techniques which will help untangle the complex internal ecology of the human cell. At any given stage in its development, a living cell may contain about 2,500 different kinds of protein, but which of these proteins is responsible for the symptoms of ageing? Senetek intends to find out.

No one expects an immediate cure for ageing, although bio-technology is obviously making inroads on some of its more unpleasant side-effects—including some of its less publicised problems, such as osteoporosis, a progressive weakening of the bones. The older we get, the less calcitonin we produce, a trend with major implications for our health, given that calcitonin appears to regulate calcium turnover in our bones.

Most victims of osteoporosis are old women, with one out of every four caucasian women over 65 showing some symptoms. The USA alone has 15 million people suffering from the disease, which results in a loss of height, deformities, and spontaneous fractures which refuse to heal and, often, leave the victim in excruciating pain. Clinical trials with calcitonin, which is already used to treat Paget's disease and hypercalcaemia, have shown promising results, and Celltech, for example, is working on the development of new products based on both calcitonin and katacalcin, which appears to play a similar role in calcium regulation.

But, however clever medical biotechnologists become in helping us to avoid such problems, there is one fact of life we shall probably never overcome: the older our skin, the more wrinkled it becomes. Many of us may find a lined face a thing of great beauty, but there will always be those who look in the mirror and be dissatisfied with what they see. Even here, biotechnology may soon have something to offer—Collagen is one US company which offers to repair your scars and wrinkles by injecting you with collagen, a white gel extracted from calf-hides. In fact collagen is the material that holds us all together, binding cell to cell and tissue to tissue. Calf-hides are not in short supply, but it is presumably only a matter of time before a genetic-engineering company announces a recombinant-collagen wrinkle-remover.

The people at whom Collagen is aiming its 'Zyderm Collagen Implant' are women in the 35–54 age-bracket with annual incomes in excess of $25,000. The company has been very careful not to over-sell the product, saying that the treatment 'can smooth certain skin imperfections—not necessarily to make you look younger, but to look better'. The last thing Collagen president Howard Palefsky wants, he says, are 'headlines reading "98-year-old woman gets 3cc of collagen and weds 28-year-old rock star"'. But dermatologists expect that there will be a good deal of interest. Said one: 'There are an awful lot of people with wrinkles out there.'

Remember, however, that this is a treatment, as Palefsky admits, 'for well people who have the option to do nothing'. There are many, many more serious diseases which are crying out for the sort of products which are being worked up in the gene factory.

SIX

The New Crops

'We are going to grow pork chops on trees.' If any one phrase brings back the early misplaced enthusiasm of the new breed of agricultural biotechnologists it is this one, attributed to Dr Martin Apple of the ill-fated International Plant Research Institute (IPRI). Apple's comment was not as stupid as it sounds: he was using a typically colourful analogy to explain to an interviewer how biotechnology would help produce new high-protein food crops, but it came back to haunt the industry when IPRI teetered, month after month, on the verge of bankruptcy, before being taken over by Bio-Rad.

The 'green revolution' of the 1960s and 1970s had transformed much of world agriculture, mankind's oldest and largest industry. Many people assume that it was simply a question of more sophisticated fertilizers and pesticides becoming available, but some of the most important advances came from plant breeders, who developed robust, higher-yielding varieties of such leading crops as rice and wheat. Like many other highly bred varieties, however, these need constant infusions of fertilizers and pesticides to support their enormous productivity and to compensate for their vulnerability to disease and pests. Now plant breeders are being issued with a new set of tools, more powerful than anything they were able to use before.

In theory, at least, genetic engineering and other biotechnologies promise to allow plant breeders to insert foreign genes into plants, enabling them to yield more calories and protein; to resist major crop diseases and flourish even in times of drought, or in salt-laden soils; to resist extremes of heat and cold; to be resistant to herbicides used to control weeds; to flourish without any need for artificial fertilizers; and, a possibility which causes a great deal of concern for tropical countries growing high-value crops, to allow plant cells to produce chemicals, such as drugs, perfumes and food additives, when grown in laboratory cultures.

Listening to Apple talk at a conference in the spring of 1982, there was no doubting his vision. 'Genetically engineered improvements in pest and disease resistance may be among the first fruits of biotechnology,' he concluded, pointing out that 'such traits are sometimes controlled by a single gene and are relatively easy to manipulate'. He spoke of IPRI research projects designed to boost the protein content of major food plants, like the potato, and of longer-term research designed to produce plants which would be very much better at photosynthesis and able to fix their own nitrogen.

IPRI had announced research contracts with such companies as Atlantic Richfield, Claeys-Luck, Eli Lilly and Davy McKee. It had set up two joint ventures with Sime Darby Berhad, the Malaysian plantation-based group; these were designed to introduce the new biotechnologies to the ASEAN region—Indonesia, Malaysia, the Philippines, Singapore and Thailand. Given that Sime Darby controlled over 200,000 acres of tropical agricultural land at the time, the news that the joint ventures would look at ways of genetically improving such crop plants as cassava, the date palm, rice and rubber was taken seriously, helping build IPRI's reputation as an industry leader.

Yet IPRI was already in severe difficulties when Apple addressed that conference. He had been manoeuvred out of the company, and others in the business, many of whom were on the verge of raising money for the same sort of work, were watching IPRI's accelerating slide towards the financial brink with undisguised horror. Even Apple had admitted that plants were likely to prove a good deal less tractable than the bacteria which the rest of the industry were inducing to perform modern miracles. 'Plants are more complex organisms,' he agreed, 'and far less is understood about plant genetics. More complete "maps" of plant chromosomes are needed, as is a clearer understanding of the relations between individual genes. Many plant functions are controlled by several genes on separate chromosomes, adding complexity to the snipping and splicing of recombinant-DNA technology.'

But he was far from chastened, stressing that 'there is a singularly encouraging aspect of plant biotechnology—the capability to regenerate whole plants from isolated cells and tissue. This feature, unique to the plant kingdom, greatly facilitates the manipulation of genetic material.' In fact, the power of some of the new plant-biotechnology techniques is quite extraordinary.

Tissue culture, for example, is roughly equivalent to taking a piece of human skin and regenerating a very considerable number of carbon-copy human beings from it. To achieve this near-miracle, the plant cells have to be duped into thinking that they are, in fact, plant embryos. This is done, or at least attempted, with bewilderingly complicated mixtures of synthetic growth hormones, vitamins, mineral salts and an energy source—typically a sugar. If this alchemy is successful the cells begin to reproduce themselves, sometimes producing millions of individual cells. These are then persuaded to form into embryos, from which complete plants can be regenerated.

Another key technique is protoplast fusion, which is increasingly used as a first step in tissue or plant-cell culture. Protoplasts are plant cells whose walls have been dissolved by an enzyme or weakened by an electric field. The power of protoplast fusion results from the fact that it is possible to persuade two protoplasts from different plants to fuse into a single cell, combining their distinct genetic characteristics. Once you have stimulated the cell's wall to reform, you can proceed to the next stage and produce a new plant—which can then be multiplied in culture.

Our newfound ability to multiply plant cells which are known to be free of disease, because of careful sterilisation prior to culturing, means that quarantine regulations are no longer needed. But there are still tremendous hurdles to be overcome. Protoplast fusion may be powerful, enabling you to engineer a plant to be tolerant to salt or drought, for example, but the process is still a surprisingly random one. Only a small proportion of protoplasts can actually be persuaded to fuse, and scientists must have the modern equivalent of 'green fingers' to achieve the plant regeneration and rooting steps. It is perhaps significant that, when Apple launched a new start-up company, Ean-Tech, with some of the cash he had received as a 'golden handshake' on leaving IPRI, he picked targets well clear of plant biotechnology.

Sunbeans and Jumping Genes

Those still involved in plant biotechnology have been noticeably less dramatic in their claims. 'No one is going to mint gold-blocks out of this next year or the year after,' said Dr Alan Robertson, a former director of ICI and now chairman of Celltech's 'country cousin', the Agricultural Genetics Company (AGC). 'It helps to understand the time-frame,' said Dr Roger Gilmour, lured back from the USA to become AGC's chief executive. 'The bit that people talk about most is the basic genetic manipulation, the transfer of genes from microbes to plants, from plants to microbes, and from plant to plant. The freedom to do that is only now emerging'.

'We've had the transfer of a gene from a microbe to a plant,' he pointed out early in 1984, 'and now we've had the transfer between plant and microbe take place, but it's not going to become an everyday occurrence. There are still major obstacles in the way. So the science isn't going to be terribly free and open for another few years. But what is known,' he stressed, 'is that it is now going to happen. It's no longer a question of whether it is possible, but of when it is going to be possible.'

When it does, however, the game will only just have begun. Gilmour warned:

Once we can transfer plant genes readily, we still have to track down the genes which are going to be of some value. We've obviously got our ideas on this—and this, in fact, is where the real competition is going to take place. No one expects plant biotechnology to be cheap. Once you have a new plant with new properties, you don't just go out and start selling it. You've got to go through the whole plant-breeding scheme, to prove that it breeds consistently, that it fits into the existing rules and regulations on plant definition, and that you can produce it in sufficient quantities to make commercial sense. The whole process could take 10 to 15 years.

But the news from around the world was exciting enough. In 1981, for example, two US scientists had announced that they had succeeded in transferring a protein gene from a bean seed to the cell of a sunflower, a member of a different plant family, dubbing the new plant tissue the 'sunbean'. Dr John Kemp, then of the Department of Agriculture's Agricultural Research Service, and Professor Tim Hall, of the University of Wisconsin, reported that the gene was not only stable in its new host, but was producing messenger RNA, a major step towards expression—and a world first.

They had used Ti (for tumour-inducing) plasmids from the bacterium *Agrobacterium tumefaciens*, which causes crown-gall disease in plants, to carry the gene into the sunflower cell, exploiting the bacterium's natural infective mechanism. 'What we did,' said Kemp, 'was to turn the bacteria's exploitation of plant cells into a tool for the transfer of genes useful to us.' Although Professor Hall admitted that 'there are a lot of steps still on the way', he suggested that the work under way in plant biotechnology was 'laying the groundwork for 21st-century agriculture'.

And another breakthrough, among many being reported, was achieved by Monsanto researchers when they inserted antibiotic-resistance genes into *A. tumefaciens* plasmids, effectively teaching the plant cells how to avoid being wiped out by antibiotics. No one pretends that such plants are under any such threat: the idea is to find a mechanism for revealing whether plant cells have taken up foreign genes. The problem had been that, while the plasmids were clearly carrying novel genes into plant cells, no one had yet been able to come up with an imported gene which was actually expressing in the host cell. The Monsanto work transformed the picture.

By tracking down the part of the plasmid responsible for turning on *Agrobacter* genes and placing their own gene right next to it, the Monsanto team succeeded in getting this 'switch' to turn on their gene, too. When the resulting cells were exposed to antibiotics, they survived, whereas normal cells died. This was a considerable step forward, but there was a possible fly in the ointment: they had not shown that the other genes transferred by the plasmid had been switched off. This was critically important because *Agrobacter*'s claim to fame (or infamy) is that its plasmid gives plants crown-gall disease, a form of cancer. Luckily, Monsanto's collaborators at the Max Planck Institute, Cologne, and at the University of Ghent, Belgium, found that some *Agrobacter* plasmids can transfer useful genes without turning the host cells cancerous.

Even more exciting, it emerged that the genes introduced by this plasmid were being passed on to the next generation of plants, through normal breeding processes. Announcing the breakthrough early in 1983, Monsanto's director of molecular biology, Dr Ernest Jaworski, predicted that it would prove to have been a vital step towards the goal of being able to transfer virtually any gene into plant cells.

A limitation of the new technique, however, was that the Ti plasmid infects only one class of plants. These, the dicotyledonous plants, include tobacco, many food plants (e.g., potatoes, beans and cabbage) and many trees. But this left such key 'monocots' as maize, rice and wheat effectively out of range. However, work sponsored at the UK's John Innes Institute by the US company Agrigenetics had shown that so-called 'gemini' viruses (or twin viruses), consisting of two strands of DNA wrapped up in twin capsules of protein, contain more free space than most other viruses investigated, space which can be used for inserting foreign genes. More to the point, they naturally infect monocot crops such as maize.

Agrigenetics, based in Colorado, had been one of the companies which had closely tracked IPRI's decline. It was not a consultancy operation: instead, it was grafting its biotechnology activities onto an existing seeds-distribution business covering more than 50 different species. This earned it annual revenues of more than $100 million, at a time when IPRI's revenues were less than $3 million (with over $10 million in expenses). Agrigenetics chairman David Padwa stressed that he had no desire to see IPRI go under, because of the damage this could do to the industry as a whole.

Agrigenetics had announced in 1983 that it had achieved gene expression in the 'sunbean'. 'This truly opens the door to the development of novel plants,' said John Kemp, by now based at Agrigenetics' Advanced Research Laboratory in Madison, Wisconsin, which was directed by Tim Hall. And Agrigenetics also linked up with Australia's CSIRO, agreeing to contribute more than $1.8 million towards work on so-called 'jumping genes' under way at CSIRO's Division of Plant Industry. CSIRO had earlier succeeded in isolating and cloning a 'jumping gene' from maize. Such genes, known as transposable DNA elements or transposons, are pieces of DNA which can spontaneously detach themselves from one site in a gene and re-insert themselves at another. They seemed obvious candidates for use as vehicles for recombinant genes.

Transposons had been discovered nearly 40 years before by Dr Barbara McClintock, but until the 1960s, when they were found also in bacteria, the scientific establishment refused to accept her pioneering work on maize. But McClintock, unlike Mendel, survived to see her work widely accepted: she won a growing number of prizes and awards, culminating in a Nobel Prize in 1983. She had originally picked maize as a simple system for study, but CSIRO was already working on wheat and Agrigenetics expected that transposons would be adapted later for use in other crops.

Despite all this excitement, however, plant biotechnologists were being rather more cautious about what they told the press: Apple's shade still haunted the industry.

'Not Chipped in Stone'

AGC, which was also launched in 1983, had had a long gestation period, aggravated by IPRI's near-collapse. 'I had to ask the question: Why did IPRI hit those problems?' Gilmour recalled later. He continued:

> And that was what really crystallised and changed the approach that investors had been taking to what became AGC.

IPRI's main problem was that they were operating as an R&D company. They were setting up a new research base and they were recruiting scientists from all over the world. The supposition was that if you put all these scientists together, it would work. But the burn rate of finance was far more rapid than the rate at which IPRI was able to come up with saleable products.

Furthermore, several of the much-vaunted joint ventures, including those with Sime Darby and Davy-McKee, had not involved any cash payments to IPRI.

AGC, by contrast, had a unique trump-card up its sleeve. Like Celltech, with its direct links to the Medical Research Council's resources, AGC has an agreement with the UK's Agricultural and Food Research Council (AFRC). This gives AGC first refusal on the R&D outputs of six of the AFRC's key institutes and laboratories working in the area of nonconventional plant breeding. Even on a world basis, this resource is in the first rank. Further, the AFRC was in the process of shifting more of its own resources into some of the areas of research in which AGC was most keenly interested. So, unlike IPRI, or even Agrigenetics, AGC was in the happy position of not having to set up a massive in-house R&D capability. Its 'critical mass' had been seen to by the AFRC. This unique facility was critical in winning the initial £700,000 needed to set up the company from the British Technology Group, Advent Technology and Advent Eurofund, Ultramar and Roger Gilmour himself.

In the early stages of its development, AGC will act as a technology-transfer company, commercialising AFRC research in three main areas: nonconventional plant breeding, microbial inoculants and biological pest-control products. The first category, in fact, is a 'catch-all' key to the really exciting areas of plant biotechnology. But these target areas, Gilmour stressed, 'are not chipped in stone'. They emerged naturally during the early stages of the company's formation and will evolve as it implements its business plan. Another bright market area is emerging in plant diagnostics (see page 231).

Meanwhile, competition is beginning to build, with AGC's major competitors including Agrigenetics, later taken over by Lubrizol, and Agracetus, the $60 million joint venture formed by Cetus and W. R. Grace. But there are still many major obstacles to overcome, with much of the basic molecular biology in the plant field still to be done.

Dr Dick Flavell of the Plant Breeding Institute, one of the institutes on which AGC can now draw, underscored the nature and scale of the problem by pointing out that, while there are hundreds of thousands of plant species and each individual plant may have up to 30,000 different genes, only about 20 plant genes had been characterised by 1983. Worse, they were not all from the same plant—and the plants involved were generally not commercially significant.

As we have seen, however, progress has been increasingly rapid in recent years, and some companies have been making headway without unravelling all the plant genetics of the plants they are developing. A key technique here is tissue culture, which is already being used fairly widely, with some unusual companies now in the running—among them Moet-Hennessy, best known for such products as Dom Perignon champagne, Napoleon cognac and Christian Dior perfumes.

Through a holding in Delbard, the largest French nursery company, Moet is involved in the test-tube production of roses. It is estimated that an initial stock of ten roses could give a theoretical yield of 2.5 billion genetically identical roses in just a year. More importantly, for a company interested in quality rather than quantity, the resulting roses reach maturity much faster than garden-grown plants bred by ordinary grafting methods, and can be made more resistant to disease. Moet also hopes to produce grapes which are resistant to the diseases that periodically ravage its vineyards.

Another company which has used tissue culture since it was founded in 1959, as a subsidiary of the Guinness brewing group, is the UK's Twyford Laboratories. Sold in 1982, Twyford succeeded in attracting funds from such investors as Biotechnology Investments and, investing for the first time outside the USA, the Plant Resources Venture Fund. Twyford has developed new tissue-culture systems and can now reproduce over 100 different species using this technique.

Exporting over 90 per cent of its products, Twyford produces a wide range of plants, from ornamentals and hardy nursery stock, through plantation crops such as the date palm and jojoba, to apple, cherry, pear and plum trees. The company is also trying to inject more science into plant biotechnology. 'There is a lot of art in micropropagation and tissue culture,' explained Dr Ken Giles, who had recently joined Twyford as research director. 'We want to reduce these activities to a scientific discipline with some special markers that we can look for to tell us how we are doing.'

The Oil-palm Revolutionaries

One company which is heavily involved in tissue culture, while keeping a much lower biotechnology profile than such US start-ups as IPRI, is Unilever. It began work on the cloning of oil palms and other crop plants, including the coconut palm, in 1968. Its development of a tissue-culture method for the propagation of selected oil palms represented the first application of the technique to a major food crop.

By the time I visited Unifield, Unilever's joint venture with Harrisons & Crosfield, in 1983, Unilever's tissue-culture work had cost it some £1.5 million. Field trials of clonal palms were taking place in such countries as Brazil, Cameroon, Colombia, Indonesia, Papua New Guinea and Zaire, but the major trials to date had been carried out in Malaysia, where a large laboratory was set up near Kuala Lumpur in 1976, with some 200 hectares (nearly 500 acres) of clonal oil palms having been planted.

Having started life in a corner of Unilever's Colworth Laboratory, Unifield produced about 10,000 cloned plants in its first year of operation, sufficient to plant about 25 trial hectares (about 60 acres). By 1983, the production target was 100,000 plants and managing director Dr Hereward Corley was talking of producing one million plants a year by the end of the decade. At the time, clonal oil palm plantlets cost £2.50 each, compared with just 14p for oil palms grown from seed, but it is estimated that there will be a need for at least 30 million plants a year by 1990, and improvements in the underlying technology, Unilever believes, will make clonal palms highly competitive.

The oil palm, *Elaeis guineensis*, provides about 15 per cent of the vegetable oil traded on world markets and is grown in plantations in South East Asia, Africa and South America, and on numerous African smallholdings. It flourishes in the humid tropics, having originated in Africa and been exported to other continents. Malaysia and Indonesia now supply the bulk of the world's palm-oil exports. Oil palms provide two distinct types of oil. First, there is palm-kernel oil, which is very similar to coconut oil and comes from the nut in the centre of the fruit; and second there is palm oil proper, which comes from the flesh surrounding the nut. The yield of palm oil far exceeds that of any other oil crop, reaching averages of 6 tonnes per hectare in well managed plantations under favourable conditions.

Today's oil-palm seeds are produced by hybridisation between a thick-shelled 'dura' mother palm and a pollen parent with shell-less fruit, the 'pisifera' type, which is often female-sterile. The resulting 'tenera' trees produce fruit with moderately thick shells and an enhanced oil yield. But, if you take a group of the best tenera progenies, you will still find striking differences in the yield and quality of oil produced by individual trees. By identifying and multiplying these élite individual palms, it is possible to create new, uniform, high-yielding clones with a performance 20–30 per cent better than today's averages.

Sadly, however, seed from tenera palms does not grow true to type. The only way to maintain the type is by cloning such plants by means of tissue culture—a technique already used with such crops as cocoa, coffee, rubber and tea. A number of these were objects of attention in IPRI's South-East Asian joint ventures, while a new joint venture announced in 1984, Plantek International, aims to develop novel strains of coffee and tea plants. It is backed by a US plant-biotechnology company (Native Plants), two Japanese companies (Kyowa Hakko and Sumitomo) and Tata Enterprises, of India.

But, until recently, there has been no suitable method for the vegetative propagation of the oil palm. It does not branch and, since it grows from a single terminal bud, it is not possible to take cuttings for propagation. Tissue-culture methods—widely used for the propagation of many horticultural plants, including orchids, lilies, ferns, chrysanthemums and strawberries—were the ideal solution. Most laboratories use shoot-tip (meristem) or 'mericloning', methods to stimulate the proliferation of buds. These are subsequently rooted and planted out. This method, however, was not successful with oil palms.

Unilever's method, by contrast, requires the intermediate development of a disorganised cellular mass known as a 'callus'. This method has the advantage that it is not necessary to kill the source palm, or 'ortet'. Instead, it is possible to start with any piece of tissue capable of growth, such as roots, young leaf-base or even flower buds. In practice, it has proved easiest to start with actively growing roots.

This root tissue is disinfected and placed in a nutrient medium where, in response to appropriate prodding with growth hormones, the cells in the tissue multiply to form the callus. If the callus forms, by no means a guaranteed step, it can be repeatedly subdivided and will grow indefinitely in culture. Once the tissues have become disorganised and are no longer identifiably root or leaf cells, they have the potential to reorganise in the form of embryo-like bodies which can develop into complete plants. So far, however, this reorganisation step is still slow and unpredictable.

The laboratory, however, is only one link in the long chain which leads to the widespread use of a new clone. The first step is the plant breeder's hybridisation programme, which creates the top-performing plants from which the very best can be selected—although you have to be very careful in picking your élite plants: a high-yielding palm may be genetically no different from a low-yield palm, simply situated in a more favourable spot. Among the characteristics which are important are the efficiency with which the palm converts sunlight into various raw materials, and converts these raw materials into fruit and oil; how much fruit it forms in each bunch; how much oil there is in this fruit; and the oil's quality in terms of composition, colour and stability. Additionally, potential ortets may be screened for disease resistance, drought tolerance or fertilizer economy. Small fronds, to permit higher planting densities, and shorter trunks, to ensure a long economic life, are other desirable characteristics.

The first clonal oil palms were field-tested in Malaysia early in 1977, bearing their first fruit late in 1978. They proved to be very uniform in their fruiting behaviour. But these first experimental clones were derived from seedlings, not from proven palms, to test cloned seedlings against the seedlings themselves. The clones have shown remarkable conformity within a clone, but distinct differences in oil composition and other characteristics between clones.

The next task was to develop clones from the very best material available from the breeding programmes, an activity which was undertaken by the Bakasawit partnership in Malaysia, the results of which are still coming in. But the overall conclusion to date is that the clonal palms will offer enormous, affordable advantages.

Test-Tube Forests

The pay-off from tissue culture at Weyerhaeueser, a company I visited in Washington State in the summer of 1981, is likely to be somewhat longer-term, although this company is just as excited about the prospects. 'If the 21st century seems closer at Weyerhaeueser than at other companies,' said company president George Weyerhaeueser, 'it is because wood fibre—the raw material of our basic business—forces us to take the long view. Through intensive management, we have nearly doubled the rate of growth of our most valuable trees. Still, the seedling-to-harvest cycle is about 60 years for a Douglas fir and about 30 years for a Southern pine.'

Convinced that world demand for timber and associated products will double by the year 2000, Weyerhaeueser has been considering ways in which it can boost the yield of its massive forest holdings. One reason I had wanted to visit the company was that it is probably the world's largest timber producer; another reason was that it was one of the earliest companies to opt for tree *farming*, instead of the cut-and-run approach operated by most of its competitors.

Weyerhaeueser has been looking at such areas as optimum tree-planting densities and thinning rates, weed competition, and the effectiveness of different cycles of fertilization and insect control. But high-yield forestry also depends on high-yield trees, so Weyerhaeueser has been selecting and breeding faster-growing trees for many years. New techniques have been introduced to accelerate these programmes, including hormone injections, designed to induce cone and pollen formation at an earlier stage in favoured trees, speeding up the generational cycle. The company's propagation programmes had been exclusively sexual up to that point, however, with seeds from superior trees used to replant cut areas.

The first generation of 'super-trees', planted in the 1950s, was expected to produce 10 per cent more timber per acre, and it was predicted that continued inbreeding could improve productivity by 50 per cent or more. 'There has to be a ceiling,' admitted Rex McCullough, a Weyerhaeueser forest geneticist, 'but we're all guessing as to how far we can go.' Various companies are collaborating in this work, to ensure that the gene base does not become too narrow. Tissue-culture experiments were also under way, although to my eye many of the resulting trees looked weedy and misshapen.

Conifer embryos were first cultured in sterile conditions as long ago as 1924, and the test-tube production of plantlets has since been reported for a number of other commercially important North American timber species, including the redwood, white spruce, longleaf pine, western hemlock and Douglas fir. But Weyerhaeueser still had a long way to go to the clonal forest —which helps explain why it formed a forestry-biotechnology joint venture with Cetus in 1983.

Chemical Attractions

Many of the companies which have been investing in plant biotechnology, however, are chemical companies, like W. R. Grace, which invested in Agracetus (see page 14). Like Grace, too, they are in the game for a variety of reasons. Some see plant biotechnology extending the markets for their existing pesticides, while others see attractions in some of the speciality products likely to emerge in the agricultural sector, including treatments designed to control frost damage of crop plants and kits for diagnosing plant diseases.

AGC, for example, is working on a growing array of such kits. 'We've got scientists in the institutes who can pick a virus off the end of a plant and tell you what it is,' said AGC's new research director, Dr Peter Dean. 'It's just what people need: the breeders need it, the farmers need it, and it's really just a question of setting up the service.' In going for such targets, AGC could well link up with a much larger company, as Celltech linked up with Boots in the human diagnostics field.

Other companies, again including Grace, are keen to see if they can produce high-value products such as flavours and fragrances from plant-cell cultures. But all are acutely aware of biotechnology's potential challenges to their existing products, whether from microbial pesticides or from bio-inoculants which, for example, could enable plants to produce their own fertilizer.

So let's take a look at three of these key areas in which major companies are showing interest: herbicide-resistant crops, microbial pesticides and microbial inoculants.

The first of these illustrates one of the two main areas in which plant biotechnologists hope to contribute to crop yields, by improving specific plant characteristics. The pesticides and inoculants are examples of the second main strategy, which involves the genetic manipulation of micro-organisms to enhance nitrogen fixation, produce pesticides, or control crop diseases and promote plant growth.

Many market analysts have forecast that herbicide-resistant crop plants will lead to a complete restructuring of the $2.4 billion-a-year US herbicide market—and where the USA goes the rest of the world can expect to follow. Among the companies already active in this area are American Cyanamid, which is funding work at Molecular Genetics on herbicide-resistant strains of corn; Ciba-Geigy, which hopes to inject new life into its atrazine herbicide, which no longer has patent protection; and Monsanto, whose glycophosphate herbicide, itself soon to lose patent cover, has featured in some of the most exciting recent plant biotechnology work.

A Way with Pests

Microbial pesticides are another area on which such companies are, at the very least, keeping an eye. Some chemical companies are already in the market with biological pest-control products. Such products exploit the natural enemies of particular pests, whether these enemies be predatory organisms, parasites or disease-promoting viruses. Kemira, for example, sells predatory mites and midges for use in greenhouses, and a couple of other biological products for use in commercial forestry. Dr Steve Lisansky, while at Tate & Lyle, which is also working on microbial insecticides and the use of *Bacillus thuringiensis* as a control agent for mosquitoes and blackfly, estimated in 1984 that western markets for microbial insecticides were worth $33–$45 million, with *B. thuringiensis* accounting for about 90 per cent of the existing market. Lisansky later set up a new company, Microbial Resources, to develop and market such products.

Trends which are forcing farmers to think rather harder about existing chemical products are their environmental effects, coupled with the various regulatory constraints designed to avert those effects, and the build-up of resistance to commonly used pesticides among soil micro-organisms. Resistance is passed from one organism to another on plasmids, those ubiquitous mobile genetic elements, in very much the same way that, in the healthcare field, resistance has built up to commonly used antibiotics.

Ultimately, though, the farmer is going to use what works, is affordable, and on the market. 'The farmer doesn't really care if it's chemicals or microbes in the jug,' as Arthur D. Little analyst David Wheat put it. But the potential market is enormous. In 1982 alone, nearly $1 billion was spent on insecticides in the USA, with the world market worth perhaps three times as much. Yet, despite the fact that they have been available since the 1940s, microbial pesticides accounted for only about $10 million of the US market, their world sales being worth perhaps twice as much.

Such products have a number of key advantages, in that their toxicity appears to be very low and they do not seem to produce resistance in the target pests. But they can be sensitive to environmental conditions, including moisture, temperature and ultraviolet radiation from the sun.

Among the new companies which have moved into this area, some, like Ecogen and Zoecon, are working on products designed to knock out even such indestructible pests as the cockroach: Zoecon is working on heart-accelerating hormones designed to give cockroaches heart-attacks!

But many see a coming shake-out in this sector of the market. 'I think the companies that will be selling the most pesticide in the year 2000,' Wheat told *Bio/Technology*'s Arthur Klausner, 'will be the same ones that are doing it now, or very similar to them.'

A Microbial Fix

Microbial inoculants are generally seen as a longer-term prospect (although some are already on sale). The role of micro-organisms in ensuring soil fertility has become increasingly clear: indeed, today's pesticides are rigorously tested to ensure that they do not damage the soil ecosystem. The contribution of such micro-organisms is perhaps clearest in the 'solid-state' fermentation process which produces one of our most highly savoured crops, the common white mushroom, *Agaricus bisporus*.

Companies like Canada's Cellex Biotechnology are intensively studying the role of these microbes, in the hope that they can boost the productivity of such fermentations. Cellex notes that only mushrooms currently convert cellulosic wastes directly into a food crop, and expects the mushroom industry to move into the production of novel products such as enzymes, anti-cancer compounds and antibiotics. This trend, in fact, has been increasingly pronounced in Japan.

'Microbial inoculants,' insists AGC's Roger Gilmour, 'are today's technology. We have a range of microbial inoculants which we are starting to market.' An example of the use of such inoculants is the success of the Welsh Plant Breeding Station in boosting the nitrogen-fixation rate of clovers. White clover can be used to help improve upland grazing, because it is a highly nutritious animal food in its own right and because it plays an important role in nitrogen fixation, through its close association with *Rhizobium* bacteria. Unfortunately, however, the rhizobia found in many hill areas are few in number or relatively poor at fixing nitrogen. By inoculating the roots of clover with improved strains of these bacteria, spectacular improvements in yields have been achieved, from less than 500 kilos of dry matter to over 2,000 kilos per hectare (450lb to 0.8 tons per acre).

Similar strain-improvement work is under way in rice-growing areas of the world, where such blue-green algae as *Anabaena azolla*, and associations between such algae and the small floating freshwater fern *Azolla pinnata*, are responsible for fixing nitrogen for rice crops in Indonesia, China, Vietnam and other tropical areas. This association, with the algae living in the fern's leaf cavities, produces floating nitrogen 'factories' which harness the sun's energy to fix atmospheric nitrogen.

In trials carried out by the International Rice Research Institute in the Philippines, 23 consecutive rice crops were harvested from otherwise unfertilized soils, with no apparent reduction in soil-nitrogen levels—thanks to blue-green algae. Peasant farmers, in a traditional form of microbial inoculation, often introduce small quantities of the algae into flooded paddies, to get the process off to a good start. The opportunities for improving this traditional technology with modern biotechnology are thought to be considerable.

In the USA Gerald Cysewski of Cyanotech has been developing ways of using such algae as the basis of new lawn-fertilizer products. Cyanotech dries the algae, without killing them, and then mixes them with the mineral vermiculite, as the carrier, and another fertilizer to supply phosphorus, potassium and trace nutrients. When applied to lawns and moistened, the algae begin to fix nitrogen. Cysewski picked the lawn market because the algae are sensitive to pesticides: domestic pesticides are generally weaker than those used in agriculture. But Cyanotech is now trying to develop algae which are more resistant to commonly used pesticides. Full-scale production could use two-hectare (five-acre) ponds in Hawaii or Nevada.

Whatever happens to such products, a sure-fire winner should be the 'microbial consortium' developed by Dr Jim Lynch of the Glasshouse Crops Research Institute and designed to break down surplus straw after the harvest. With tightening controls on straw- and stubble-burning, the only alternative has been to plough surplus straw back into the soil. But this typically brings significant yield penalties in later plantings, because of the toxins released into the soil during the breakdown of the straw. Lynch's new bio-process, developed at the Letcombe Laboratory, involves the use of a mixture of micro-organisms, including a fungus which thrives on cellulose and a nitrogen-fixing bacterium. It worked under laboratory conditions, so AGC gave the AFRC a £250,000 contract to bring the process up to speed for field trials in the autumn of 1985.

Fertilized By the Sky

The ultimate objective in nitrogen fixation, however, is to use genetic engineering to introduce nitrogen-fixation genes directly into crop plants. To get an idea of progress to date, I visited the Agricultural Research Council's Unit of Nitrogen Fixation, at Sussex University, Brighton. Its director, Professor John Postgate, is convinced that nitrogen fixation remains a major problem for the future, and that biological methods are our best hope of finding a long-term solution. He suggested that the energy and transport costs associated with the industrial Haber-Bosch process of nitrogen fixation, which requires high-pressure, high-temperature equipment, make it an unattractive long-term solution to the crop-fertilizer needs of the world, as do the problems associated with fertilizer run-off and the build-up of nitrate pollution in rivers and water supplies. Increased use of nitrogenous fertilizers is inevitable, however, as is the increased use of existing nitrogen-fixation systems. But neither, Postgate argued, will be equal to the task of feeding the extra billions of people the planet will have to support in the 21st century.

Founded in 1963, the unit's original remit was to investigate how biological nitrogen-fixation systems worked. Then, in 1971, it attracted a good deal of publicity when it succeeded in transferring *nif* genes, which confer the nitrogen-fixing ability, from *Klebsiella pneumoniae* bacteria to *E. coli*, which had not previously enjoyed this ability to pluck nitrogen from the air.

The 17 *nif* genes, Postgate explained, represent 'the biggest package of genetic information that is not itself a virus or an autonomous entity like a plasmid'. This cluster of genes has since become one of the most intensively mapped genetic systems in molecular biology. But, warned one of the unit's molecular geneticists, Dr Mike Merrick, 'there is a great deal to be done before we achieve new systems of immediate practical benefit'.

Although one example of nitrogen fixation has been reported in a eukaryotic microbe in Japan, it has been generally accepted that it is exclusively found in prokaryotic microbes in oxygen-free conditions. This fact hints at the problem genetic engineers face in their attempts to transfer working *nif* genes into such eukaryotic organisms as crop plants. The unit has switched much of its work into *Azotobacter* and hopes to obtain *nif* gene expression in a plant chloroplast—the most prokaryotic environment available in a plant cell. The implications of ultimate success are 'immense', Postgate stressed, but so, it seems, are the obstacles along the road to the self-fertilizing crop of the future.

Seeds of the Future

'Seed is the delivery system for the products of the new plant genetics,' says one industry analyst. Undoubtedly, most of the most important near-future contributions which biotechnology will make to plant agriculture will be in the field of primary products, particularly in the improvement of seeds and seed proteins. And this is why there has been a Klondike-style rush to buy up seed companies, with over 100 being bought up in the USA alone.

Most seed companies are relatively small: nearly all the members of the American Seed Trade Association report sales in the range $3–$10 million. Yet this scale of operation will no longer be sufficient to support the sort of biotechnology work which is becoming part of the entry fee to the industry. 'The game of molecular biology,' explained Agrigenetics chairman David Padwa, 'is not a game for three people over a garage.'

Hybrid seeds, which are produced by cross-breeding two different varieties of the same plant, have excited enormous interest, not least because they do not breed true—which means that farmers cannot save this year's seed to plant next year: they have to go back to the seed company to buy more. The hybridisation process also gives companies something of a stranglehold on their products: they can keep the identity of the parent strains secret. Moreover, because seed generally accounts for less than five per cent of a farmer's total costs, the industry believes that it can charge a premium for hybrid seeds. Hybrid corn and hybrid sorghum, like the new hybrid wheat strains, have also proved to be more resistant to drought, cold and disease.

But, even though the new hybrid wheats can increase yields by up to 25 per cent, farmers are not yet totally convinced, largely because hybrid seed can cost more than three times as much as normal seed. Meanwhile, biotechnology is not alone in the field. In the late 1970s, Rohm & Haas developed a chemical hybridisation approach, which involves spraying the crop you want to hybridise with a chemical that prevents it self-pollinating, but leaves it wide open to cross-pollination. Companies like Monsanto and Shell are also now using this technique.

Fragrant, Flavoursome Cells

Not surprisingly, the political implications of these trends are causing some concern, with many observers extremely worried that the larger companies will simply crush their smaller competitors. There are also concerns about the genetic implications of many of the current trends in plant biotechnology, as we shall see in Chapter 10. But one of the most immediate fears, at least as far as the Third World is concerned, is that the developed countries will soon be growing all the high-value crops they want in laboratory fermenters rather than import them from economies which are totally dependent on them.

They will not have been reassured by the news that the Japanese, in the form of Mitsui Petrochemical Industries, have worked out a way of producing one extremely expensive plant product, the drug *shikonin*, by plant-cell culture. Natural *shikonin*, extracted from the roots of the shikon plant, costs about $4,500 a kilo, perhaps a thousand times as much as ordinary foodstuffs. The shikon plant takes three to five years to grow to the stage where its roots contain a maximum of two per cent of the drug, used to treat inflammations and kill bacteria. Yet, by selecting the most productive cells from tissue cultures, Mitsui found it could produce 15 per cent concentrations of *shikonin* in very short order.

Like many of the flavourings, fragrances and other speciality chemicals which biotechnologists may one day extract from plant cells, *shikonin* is a secondary plant product, a 'secondary metabolite', so called because their function is often difficult to discern and they are generally molecules which are made by proteins, rather than by the genes which produce the proteins. They are typically of low molecular weight and often structurally complex. They include aromas, drugs, emulsifying agents, enzymes, flavours and insecticides.

To get some sense of the prospects, I visited Sheffield University's Wolfson Institute of Biotechnology, where Professor Mike Fowler had launched a new company, Plant Science, in 1982. Plant-cell culture, he noted, is 'very much an enabling technology, allowing the further exploitation of the plant kingdom as a speciality chemicals resource'. Plants, he pointed out, are 'unrivalled in their synthetic capabilities and in the range of chemical structures they contain'. The total number of known structures is in excess of 20,000, and the annual rate of new discoveries is running at about 1,600. Secondary metabolites which are already used include codeine, digoxin (an analgesic), diosgenin (an anti-fertility agent), morphine and quinine.

The new company is seen as a means of transferring the relevant technologies from the Wolfson Institute into industry. But, Fowler stressed, there are still many problems in getting plant-cell cultures to go in the direction you want them to go. Plant cells differ from microbial or animal cells in a number of important ways: they are much larger than microbial cells; they have a limiting cellulose cell wall (which has a high tensile strength and is fairly plastic—but is vulnerable to shear forces in a fermenter); and they tend to form internal spaces, or 'vacuoles', in which toxins can collect. They also have much slower cell-division rates than microbial cells.

New techniques, such as fine-cell suspensions, are being devised to gain greater control over the growth and metabolism of such cells, but a degree of clumping may actually prove to be necessary if secondary-metabolite production is to take place at all. Fowler concluded by saying that 'plant-cell culture is not the universal panacea of the speciality-chemicals sector', but expressed the conviction that there are 'some extremely good product targets'. And, as new processes are developed and experience is gained, 'the product window will undoubtedly widen. The tremendous range of chemical structures present in plants almost guarantees major input from the plant kingdom into the chemical industry'.

So, in conclusion, no universal panaceas and no pork chops on trees, but plant biotechnologists now have at their disposal a cluster of technologies which promise (or threaten, depending on your viewpoint) to trigger a second Green Revolution. This revolution, however, will proceed on a much broader front and, on the evidence presented so far, could help solve many of the problems thrown up by the last great agricultural convulsion.

SEVEN

The New Breeds

The Arab horse was once the single most important yardstick for measuring a Bedouin's wealth and honour. Over the centuries, the Arab breeders built on the foundations provided by genes coding for a remarkable animal, able to live on next to nothing as it carried warriors on raids under a sun which sent most other life-forms scurrying beneath the dunes. And, even if the pure-bred Arab horse (*kehailan* in Arabic) no longer plays such a central role in such countries as Iraq, Jordan, Saudi Arabia and Syria, there are those who have been unable to shake off the breeding habit.

'Up to my grandfather's generation the Bedouin still had the people who knew about horses,' said Princess Alia, responsible for the running of the Royal Jordanian State Stud, on the outskirts of the country's capital Amman. 'Then the younger ones lost interest or went in for racing. But we realise that the Arab horse is an important part of our culture. It's a matter of honour to have correct pedigrees.' And money: US buyers will gladly pay well over $50,000 for a yearling Arab filly.

A fair amount of anybody's money, yet it pales into insignificance compared with the £233,000 paid by the UK's Premier Breeders for a black-and-white Holstein bull calf called 'Pickland Elevation B-ET'—but known informally as 'Pickles'. Why so expensive? Again, it was a question of breeding. His mother, Hanover Hill Barb, produced over 10,000 kilos (about 2,200 gallons) of milk in 1981, with nearly 5 per cent butterfat. And his father, 'Round Oak Rag Apple Elevation', was described by Premier as 'probably the best known sire in artificial insemination of all time'.

If Pickles takes after his father, he will produce something like 40,000 phials of semen per year. These will be frozen and sold to cattle breeders around the world at around £50 a phial. Yet, even today, sexual reproduction is a game of chance. Because the sperm and ova of any two parents contain only a random selection of their genes, the total number of possible genetic combinations borders on the infinite. So breeders are doing everything they can to tilt this game of chance in their own favour, like the gambler who loads the roulette wheel.

Herd in a Suitcase

This quest is leading breeders in directions which sound more like science fiction than science fact. In 1983, for example, a herd of 500 pedigree Friesians left England for Egypt in four sealed flasks no bigger than suitcases. They travelled as seven-day-old frozen embryos in flasks of liquid nitrogen, weighing in at 45 kilos (100 lb) instead of the 175 tonnes they would have weighed as young heifers. Removed from their mothers by the newly formed UK company International Embryos, in an operation taking less than ten minutes, the embryos were to be re-implanted in Egyptian cows.

The techniques of embryo transfer are fairly well known— indeed the 'ET' part of Pickles' name denoted the fact that he was a product of such a transfer. A breeder induces a pedigree cow to 'superovulate', producing a number of eggs; then the eggs are artificially fertilized by the semen of a top-quality bull. The fertilized eggs are flushed from the cow when they each contain, at most, a few dozen cells. Frozen in liquid nitrogen at a temperature of $-197°C$ $(-323°F)$, they can be transferred to local cows and, following an otherwise normal development process, are born in the environment in which they are to be reared.

This process has many advantages, and has been attracting a good many investors. The transferred animals are spared the shock of acclimatising to a new environment as calves, and also obtain from their foster mothers antibodies which protect them against local diseases; the frozen embryos may be stored almost indefinitely, and the average selling price for a pedigree embryo in the early 1980s was around £250. But there are still risks. 'Don't rely on just one cow,' one investor warned. He had bought a $300,000 cow a few years earlier which had still to produce a single embryo.

Embryo transfer was first reported in 1890 by Walter Heape, after a litter of rabbits, transferred as embryos, was born in his laboratory. Much later, in the 1930s, Heape was involved in work which resulted in the successful birth of a lamb following embryo transfer and, in 1933, his group reported the first interspecies transfer, involving sheep and goats. By 1984, scientists were reporting the birth of a zebra which had been implanted in a surrogate mare, the first time a horse had given birth to a zebra. The first calf born following such a transfer came in 1951, and the process is now used to speed the improvement of a growing number of herds.

Typically, a low-grade animal can be raised to a top-quality beast in just one generation, compared with four to five generations when traditional AI methods are used. In ordinary circumstances, exceptional cows have a limited breeding life and may produce no more than three or four calves in a full breeding lifetime, whereas with embryo transfers a single female may produce ten or more calves a year. Another key advantage is that élite pedigree embryos can be transferred to non-pedigree foster mothers, with the resulting progeny being acceptable for entry into pedigree stud registers.

Companies like International Embryos have been working also on new techniques which permit the cloning of embryos, producing identical offspring. At first it was thought that you could only split an embryo in half if you wanted the resulting calves to survive, but Louisiana State University reported in 1984 that it had achieved another first: a calf called 'Two Bits', born to a surrogate mother, had resulted from just a quarter of an embryo. A second calf from this embryo division was born to another foster mother just 24 hours later.

And work done by Dr Steen Willadsen and Carol Fehilly of Britain's Institute of Animal Physiology had already shown that 'scrambled' calves can be produced combining the genes of *four* parents. This was achieved by fusing parts of two different embryos, prior to re-implantation. When four calves were born, having been carried in pairs in two foster mothers, one proved to be pure Friesian, one pure Hereford, and two were 'chimaeras' containing a mixture of genes from both sets of parents.

Among the characteristics likely to be favoured by breeders using such techniques are improvements in growth rates, milk production, carcass lean-meat content and disease resistance. And it has also become possible to determine the sex of a six-day embryo, following a breakthrough by a US start-up company, Genetic Engineering. This development, based on the use of monoclonal antibodies, is critical, since normally breeders have to transfer embryos without the slightest inkling of whether they are feeding the surrogate mother for nine months to produce a female calf worth perhaps $2,500 or a male worth just $50 as veal. Now, with a technique reportedly better than 90 per cent accurate, dairy breeders should be able to transfer mainly female embryos, while breeders interested in meat production can concentrate on male embyros. For various reasons, however, it will be some time before such techniques are widely used.

Tigrons and Mighty Mice

Many of us imagine that evolution is an inordinately slow process, the genetic equivalent of the grinding, millimetre-by-millimetre progress of a glacier. Yet the continuing evidence of evolution is all about us.

Take the rabbit. Practically wiped out in the UK between 1953 and 1955 by the introduced myxomatosis virus, rabbits were thought to have been effectively removed as an agricultural pest. Only one per cent of the country's wild rabbits are thought to have survived that first pestilence, but just 30 years later scientists showed that 50 per cent of wild rabbits injected with the virus either failed to develop symptoms or survived the disease. The signs are that the rabbit, in a very short period of time, has become genetically resistant to the disease.

By contrast, we have had thousands of years to reshape the genetic make-up of the species we have domesticated and whose breeding we have increasingly controlled. Indeed, take a close look at some of the animals in factory farms today, and you may find yourself wondering whether you are looking at animals or biological machines. In effect, you are looking at both.

The dairy cow, for example, has been remorselessly bred to produce more milk. To try to maintain some sort of balance between her milk output and her food intake, she is fed high-energy concentrates which often give her acute acid indigestion and make her vulnerable to such illnesses as milk fever and mastitis. Her diet, coupled with all that standing around on concrete, also tends to make her lame. As a biological machine, a farmer might think, she could be better designed.

As far as meat-making machines are concerned, the pig has few rivals. Today's intensively reared pig lives indoors for its entire, brief life, being fed on a computer-formulated diet of cornmeal or soybean meal, supplemented with proteins, minerals and vitamins. Unless it has been marked down for breeding, it goes to the slaughterhouse at the age of five or six months. But if the pig thought human ingenuity had reached its limits as far as porcine redesign was concerned, it reckoned without genetic engineering.

No one who has been reading or watching the news during the 1980s can have missed the fact that genetic engineers are beginning to do some rather strange things with animals. Among the simpler announcements, the Vicomte Paul de la Panouse reported that one of his ligrons (the offspring of a lion and a tigress) had produced a cub, a tigron—although it was not clear who, in a park where the animals are free to roam and mate at their leisure, had fathered the beast. 'This is the first time that a hybrid such as a ligron has given birth,' said the Vicomte, 'thus disproving the theory held until now that hybrids are sterile.' This general 'rule' had been derived from experience with mules, which result from the crossing of horses and donkeys, but later in 1984 a Nebraska farm reported the first authenticated delivery of a foal by a mule.

But an even stranger announcement had been made, just a few months earlier, by the Institute of Animal Physiology. It had produced a sheep/goat chimaera which behaved like a normal goat, had goat-like horns twisted around in a sheep-like fashion, sported long wavy goat wool, and proved to have both sheep and goat red cells in its bloodstream. Unlike the sperm of the mates of the ligron and Nebraska mule, however, the goat-sheep's sperm turned out to be defective, so that it proved infertile in natural matings with female goats.

Not to be outdone, scientists at the University of California at Berkeley announced that they had cloned DNA from preserved tissue of the quagga, an extinct animal related to the horse and zebra. Although the quagga had been extinct for more than 100 years, they tracked down specimens of the animal's skin in the Museum of Natural History at Mainz in West Germany. Extracting DNA from dried muscle tissue, they introduced it into bacteria to produce more. But they were quick to point out that they had been able to retrieve only a few of the millions of genes which would be needed to reconstruct a complete animal, so the reassembly of a mammoth or dinosaur will almost certainly remain the stuff of science fiction.

Those Californian scientists were unconcerned that their research had no commercial implications. By contrast, the scientists who injected rat growth hormone genes into fertilized mouse eggs, which were then implanted in foster mother mice, must have had more than a sneaking suspicion that a successful outcome would have commercial implications. While people are unlikely to be attracted by mouse-meat, the fact that some of the resulting 'super-mice' stimulated by the rat hormone grew to nearly twice their normal size has momentous implications for animal breeding.

Equally exciting for those involved, some of the super-mice passed on the implanted genes to their own litters, and the same research team later produced similar super-mice by injecting human growth hormone genes. But this process of direct micro-injection of DNA into mammalian eggs is still a hit-or-miss affair at best. Only about a third of the mouse embryos injected with the rat genes proved to have incorporated them—although even this result was a radical improvement on earlier outcomes.

As far as the commercial implications of the work are concerned, it opens up the prospect of giant pigs, sheep and cattle, perhaps capable of producing much larger quantities of meat or milk. 'If we can make bigger mice,' as Dr Ralph Brinster of the University of Pennsylvania put it, 'we can make bigger cows.'

The super-mice were the end-result of years of work by three scientists: Dr Richard Palmiter, of the Howard Hughes Medical Institute at the University of Washington, Seattle, and Drs Ronald Evans and Neal Birnberg of the Salk Institute in La Jolla, California. The gene they inserted was a clever amalgam of the rat growth hormone gene with an 'on-switch', or promoter, snipped out of a mouse gene. The mouse portion of the artificial gene switched on the rat portion. The composite gene was inserted into the mouse embryos at the University of Pennsylvania's School of Veterinary Medicine.

And this research also opens up another option. The extraordinary levels of rat growth hormone produced in the super-mice started some scientists thinking about the prospect for 'genetic farming', with animals used to produce large quantities of medically useful products. Bacteria, yeasts and other micro-organisms have proved highly adaptable vehicles for genetic-engineering experiments, but they often fail to produce exactly the molecule we need—and sometimes they may be unable to produce a desired molecule in economic quantities, if at all. The possibility now is that the harvesting of such substances from large animals may prove more economic.

Recombinant Animal Farm

Down on Orwell's Animal Farm, the pigs were more equal than the other animals, so let's look at the ways in which genetic engineers may help produce the super-pig of the future. It is 40 years since Orwell's pigs burst on the world, and they would hardly recognise the pig industry of the 1980s. Pig farmers use a welter of modern technology to ensure that their meat machines produce at full throttle.

Pig breeders now use the latest population-genetics techniques in their crossbreeding programmes. Crossbred pigs, produced from two to four purebred breeds, are often superior to their purebred parents in terms of their rate of survival, growth rate, efficiency of feed conversion into meat, and number of offspring. Pig farmers also increasingly use highly sophisticated computer modelling techniques in their attempts to squeeze pig physiognomy to its very limits. Often, too, ultrasonic devices are used to scan the pigs and determine how much meat they are carrying.

Gene-splicing techniques, however, represent a major new set of tools, both for animal breeders wanting to produce more meat, more milk or more eggs more rapidly, and for companies which are developing products designed to keep intensively farmed animals healthy. Again, let's consider some of the implications for the pig.

Ultimately, perhaps, it may be possible to use direct gene transplants to produce the porcine equivalent of super-mice. Researchers at Iowa State University, for example, are trying to identify and isolate the genes responsible for conferring disease resistance or reproductive prowess in pigs. They have found animals with different levels of immune response to the disease pseudorabies, caused by a herpes-like virus. If a gene is shown to be responsible for these differences, the next step will be to track it down and clone it, prior to transferring it into pig germ-cells or early-stage embryos.

A rather more extraordinary prospect is that recombinant-DNA technology may also soon be used to redesign the microflora inhabiting the pig's digestive tract. A key target here would be to enable a pig to produce the enzyme cellulase, which would permit it to extract nourishment from cellulose, the most abundant constituent of plants. Given that the pig's digestive system is remarkably similar to ours, with the result that pigs often compete for the same basic foods, this would be a tremendous step forward for pig farmers—whatever pigs may think of the idea.

But, for the moment, the large-scale production of pig growth hormone, and possibly amino acids for use in pig feed, by genetically engineered bacteria seems to be the most imminent outcome. Synthetic growth hormone is also being investigated for any effects on the growth of lean tissue in a range of animals, including pigs. In view of the heated controversies which have raged around the use of hormones to boost animal weight, this application of genetic engineering is worth looking at a bit more closely.

The Fatted Calf

Olympic athletes undergo rigorous testing to check whether they have been taking steroids, to beef up their bodies and boost their performance: many have, although if they have been clever it can be difficult to prove. But a far more contentious application of steroids and other synthetic hormones is in the fattening of farm animals.

Produced largely by an animal's ovaries or testes, it was found that such steroid hormones not only affect an animal's sexual development and behaviour but also boost its growth. Used to enhance the growth of calves, these hormones can increase veal production by 7–8 per cent, aside from their use to fatten castrated steers and cows marked down for slaughter after years of milk production.

Natural hormones, which have to be extracted from the appropriate animals, have been scarce and expensive. So the farmers have switched to synthetic hormones. Then evidence began to emerge suggesting that some of these synthetic growth hormones were having highly undesirable effects on some of those who had eaten meat from treated animals. In 1980, for example, the news broke that one of these hormones, a stilbene called diethylstilboestrol (DES), was getting into baby food in Italy by way of cheaper cuts of veal and causing the infant cancers and appearance of inappropriate sexual characteristics in children for which it had already become notorious.

The Bureau of European Consumers' Unions (BEUC) campaigned hard for a ban not only on DES, but on all added hormones. In the event, the EEC banned only stilbenes, setting up an expert group under the chairmanship of Professor Eric Lamming to review the evidence on the effects of hormones in general. Reporting early in 1983, this group cleared three natural hormones: oestradiol, progesterone and testosterone. But it judged that more evidence would be needed before a decision could be made on two synthetic hormones: trenbolone, which mimics the effect of testosterone, and zeranol, which mimics oestradiol.

The various EEC countries were left to make up their own minds. The 'do-nothing' camp comprised the UK, Ireland and Luxembourg: these countries continued to permit the use of all five hormones. West Germany, rather more cautiously, allowed only the natural ones, while everyone else banned all five for fattening purposes. When the European Commission later came under pressure from such hormone producers as the US International Minerals and France's Roussel Uclaf, which wanted the bans lifted, BEUC maintained that both synthetics were under suspicion of causing cancer or other undesirable effects in laboratory animals. The companies retorted that such hormones could well have such effects at unreasonably high doses, but that any differences between natural and synthetic versions had yet to be proved. They also pointed out that if the hormones are properly administered, in many cases by injection into an animal's ear rather than into muscle tissue which people are going to eat, the levels of hormones left after treatment would be very low: they suggested a comparison of 0.3 parts per billion (ppb) in treated meat, against 0.2 ppb in untreated meat. The industry also warned that, if a ban was enforced, many farmers would continue to track down suitable hormones, having found that they could add meat worth perhaps £25 to a steer carcass for a very small outlay.

Whatever the ethics of the situation, some biotechnology companies, meanwhile, thought they had an answer to the problem. Genentech was just one of the companies working on bovine growth hormone, a non-steroid growth factor. But the initial results in field trials were not entirely promising: bovine growth hormone, the results suggested, might be used to boost milk production or alter the ratio between lean and fat tissue, but it did not appear to be having much effect on the ultimate weight of the carcass—which is the basis on which the market pays farmers.

A UK discovery, first announced in 1981, may hold the key to a more effective approach. The discovery was that, if you immunise animals against their own somatostatin, which controls the natural production and release of growth hormones, the brakes on their growth are released: in the case of experimental sheep, they can grow up to twice as fast. Stuart Spencer and Diane Williamson of the AFRC Meat Research Institute in Bristol had welded together a human blood protein and synthetic somatostatin and, by injecting it into sheep, ensured that the animals produced antibodies against the somatostatin produced by their own hypothalamus. Spencer stressed that the resulting response was much more natural than when you simply inject a single hormone, because somatostatin also controls the release of such other hormones as insulin, together with the intestinal and thyroid hormones.

This work suggests the longer-term possibility that self-immunisation could be used to control stress, appetite, digestion, fat production, reproduction and milk production in farm animals. By then, it seems, farm animals will already have been treated with a growing array of veterinary products generated by recombinant-DNA technology.

The Hi-Tech Vet

When the vet arrives on the farm of the future, if indeed such on-site calls are needed by then, his armoury will be very much more powerful than it is today. One of the reasons that some of the biotechnology companies have been moving into the animal healthcare field is that such products will have, as Genentech explains, 'shorter routes to market than are typical of human pharmaceuticals. Because they can be tested directly in the target animals, studies for safety and efficacy leading to regulatory approval may be less time-consuming.'

One product Genentech has been working on is bovine interferon, a naturally occurring protein in the immune systems of cattle and the first animal interferon to be produced by recombinant-DNA technology. If successful, it will fill an urgent and long-recognised need in the cattle industry. A major cause of cattle loss in the USA is a viral disease called 'shipping fever'. Some 20–30 per cent of all cattle being shipped around the USA contract the fever, and the cost of the resulting severe weight losses or deaths is reckoned to add up to at least $250 million a year.

Another novel product which vets began to offer to North American farmers in 1983 is a monoclonal antibody designed to protect newborn animals against scours. Produced by Molecular Genetics, Genecol 99, which protects newborn calves from scours, helped the company to its first full-year profit in 1983. Pathological bacteria, from the K-99 strain of E. coli, frequently infect cattle and pigs during their first day of life and can cause death from dehydration. Several companies, including Cetus Madison (later wrapped into Agracetus), had already marketed recombinant vaccines against scours. These are given to pregnant animals, so that they can transfer protective antibodies to their young; Cetus vaccines are available to protect both piglets and calves. But, according to Molecular Genetics, farmers 'have not readily accepted the complicated task of prenatal vaccination. Our staff saw a new approach. Why not design a product to give directly to the newborn?'

Produced either in mice or by tissue culture, Genecol 99 is given orally and combines with the K-99 bacteria in the animal's stomach, thus preventing infection. Molecular Genetics has estimated that in the USA alone more than one million calves die from scours each year, with several million animals requiring extensive (and expensive) treatment with antibiotics and special rehydration fluids.

In fact, most of the recombinant-DNA vaccines, monoclonal

antibodies, interferons and similar products now being developed for human use have their animal counterparts. And, as is the case with vaccines for human use, animal vaccines offer considerable safety advantages. Many major outbreaks of foot-and-mouth (hoof-and-mouth) disease, for example, are thought to have been triggered by defective vaccines. Genetic engineering will mean that there is no danger of live viruses being injected with the vaccine.

The US Animal Health Institute estimated that the 1983 market for veterinary biological products was worth only $166 million. Ignoring animal growth hormones, *Genetic Technology News* forecast that the US market for new veterinary markets would rise to $300–$400 million a year late in the 1980s. *GTN* pointed out also that

> unlike an antibiotic, a biological veterinary product can often be used against only one disease and often for only one animal species. This means that few recombinant-DNA vaccine products will find markets exceeding $5 or $10 million per year. Vaccines are generally difficult to patent, so competition is severe and the prices manufacturers can get are low. Most vaccines sell for less than $1.00 per shot at the manufacturers' level.

By contrast, the pet market may be less price-sensitive.

The animal interferons and lymphokines that are beginning to emerge offer a hope of preventing or curing viral diseases for which no treatment currently exists. Equally important, they can be used for more than one disease and for many species, ensuring that they will find a more valuable market than many products aimed at livestock.

The potential markets in the animal health care field are sufficiently attractive, however, to ensure continuing—indeed growing—interest in the biotechnology industry. The key factor will probably be the lesser product-testing requirements. But there are longer-term challenges for any biotechnologist game enough to go for them . . . and with the resources to underwrite the exercise. These targets include the self-shearing sheep and fish designed to thrive in polluted waters.

Goodbye to Shearers and Silkworms?

The self-shearing sheep is on its way. Scientists at the Commonwealth Scientific and Industrial Research Organization (CSIRO) have identified a protein which, when injected into Merino sheep, causes fleece to simply fall or peel off. There are few, if any, significant side-effects. This protein, called epidermal growth factor, is extracted from the salivary glands of male mice. Clearly, it is currently in short supply, but CSIRO hopes either to extract it from cow's milk or to produce it using genetic engineering.

Genetic engineering will certainly help farmers boost the productivity of existing farm operations, but it may also eventually bring totally new animals onto the farm. This does not mean that centaur cutlets will turn up on the menu: more likely, we shall see some better known species domesticated or further modified. Aquaculture, or the farming of fish, shellfish, crustaceans and marine algae, could well benefit from some of the new techniques.

Take salmon. Companies like Weyerhaeueser have been investing in early forms of ocean ranching, with salmon raised to the parr stage and then released out into the Pacific. After several years, the idea is that the mature salmon, fed at Nature's expense, will home back in on the installation from which they were first released. But when I visited Oregon Aqua Foods, a company in which Weyerhaeueser had invested, less than one per cent of the salmon released were returning. Yet it would take only a few more per cent to return to make the exercise profitable. So a key activity has been research designed to uncover any biological or other barriers to such ocean ranching.

If the cause of these low returns proves to be something like an inability of the farm-reared parr to acclimatise fast enough to the salinity or any other feature of their ocean range, it may be that genetic engineers will be able to transfer genes into their eggs which will confer a degree of resistance. A striking characteristic of the salmon egg is its size, and this would make the microinjection of DNA that much easier.

Salmon are notoriously fastidious about the quality of the water in which they swim but, if recent work with trout is anything to go by, perhaps salmon might one day be redesigned to be less fussy. Many of us might think that it would be a far better thing if we devoted our energy and ingenuity to cleaning up the environment, rather than to re-engineering wildlife so that it can survive in the world we are creating, but this work on trout at least hints at genetic engineering's potential for bringing new species onto the farm.

If all goes well, the super-trout which are on the genetic drawing board at the University of Southampton will incorporate genes from mice and frogs. The basic idea is to produce fish which grow faster, reach a larger size and can flourish in conditions which would kill most normal trout. A research team led by Dr Norman Maclean is hoping to transfer three separate genes, each of which has already been isolated and cloned, into rainbow-trout eggs. The first gene, extracted from mice, confers resistance to poisoning by heavy-metal pollution. Since the most damaging aspect of acid-rain pollution, that bane of northerly countries located downwind of major concentrations of industry, is not the acidity itself so much as the heavy metals that the acid leaches out of rock and soils, a successful result might mean that super-trout could be re-introduced into lakes and rivers which have become ghost waters. 'We're pretty sure the gene will express itself in trout,' said Maclean. 'But to be of any use it will also have to be integrated into the germ line so that it gets passed on to the next generation.'

The same holds true of the other two genes. One, extracted from frogs, makes globin, a component of haemoglobin, and could help boost the respiratory efficiency of the fish; this, in turn, could allow them to thrive in warmer waters than normal, where oxygen levels are lower. Salmon, which are acutely sensitive to low oxygen levels, might be similarly 'improved'. The third gene, which also comes from mice, controls the production of growth hormone. Because the language of DNA is common to all species, super-mice are an important milestone on the road to a broad range of super-species, including such super-fish.

The Chinese, according to Maclean, are working hard on the genetic re-engineering of carp, along similar lines. 'They seem to. be going all out for the growth hormone gene,' he noted. China, however, is also aware that one of its key industries is threatened by genetic engineering. If present trends continue, the silkworm may one day end up in the world's museums alongside the abacus, horse and sheep-shearer's clippers.

An estimated ten million farmers currently produce silk in China, with another 500,000 workers engaged in the production of various silk fabrics. *Bombyx mori*, the most commonly farmed silkworm, says Jean-Pierre Garel of the Université Claude Bernard, in Lyon, 'is the only fully domesticated insect whose survival depends entirely on being bred and reared by humans. The main trend in 4,000 years of sericulture has been to increase the quality and quantity of silk by improving the productivity of *B. mori* strains.'

Garel has been promoting the use of the silkworm to replace *E. coli* and the fruit fly *Drosophila* as model species for research on developmental biology. There is a considerable base of breeding knowledge to build on: China boasts over 300 varieties of *B.*

mori, while Japan has a stock of some 320 mutant strains. Any breakthroughs in such research would also be likely to have fairly immediate commercial applications.

Biotechnology has already made major contributions to sericulture: Louis Pasteur helped pinpoint the cause of a deadly silkworm disease which was decimating France's silk industry in the 1850s. Almost certainly, too, biotechnology will bring major advances in silkworm breeding. Extraordinary breeding experiments have been tried before, however, at least one of which was an ecological disaster.

An American living in Massachusetts, Leopold Trouvelot, attempted to interbreed silkworms with gypsy moths in 1869, hoping that the resulting creature would feed on oak rather than mulberry leaves but nevertheless produce silk. Unfortunately, the cage in which he kept his gypsy moths was smashed in a storm and they escaped. They reproduced prolifically and, after more than a century as an increasingly significant pest, caused massive defoliation damage to more than eight million acres (3¼ million hectares) of trees in the north-eastern states during 1982 —sparing only the tulip poplar and dogwood.

But there is a far more ominous possibility, at least as far as the silkworm industry is concerned, suggested by the fact that genetic engineers have already cloned the gene for the giant silk protein. So far, this has been only as part of a basic research project. But, given enough time, perhaps an industrial process could be devised to replace the silkworm?

If they ever become redundant as silk-producers, silkworms may still find a less romantic place on the future farm: the Chinese already eat them, stir-fried, as an apparently appetising protein supplement to a largely vegetarian diet.

EIGHT
The New Menus

'Scotland's answer to Japanese whisky,' the advertisement began. 'Tartan soy sauce. A natural product that tastes as good as the Japanese equivalent. But takes four weeks to ferment instead of three years—and is very competitively priced.' The world's fastest-fermenting soy sauce was developed by Dr Brian Wood, of the biosciences department at Strathclyde University, and was first produced in 1982 by Bean Products of Cumbernauld.

How did the Japanese take this challenge to one of their most traditional food industries? I visited Kikkoman, Japan's largest brewer of soy sauce, or *shoyu*, shortly after the Scottish company announced its new product.

Kikkoman, based in Noda City, boasts that 'there is no food-store, supermarket or department store in Japan that does not carry Kikkoman soy sauce', and it prides itself on the quality of its product. The average Japanese gets through an astonishing ten litres of *shoyu* each year, but this does not mean that the Japanese palate is undiscriminating as far as the quality is concerned. Indeed, the Japanese would echo the Scottish claim that Japanese whisky lacks the subtlety of the original by insisting that a *shoyu* product produced in days can never hope to rival the subtlety of one matured over many months.

Early forms of such fermented foods, some of which give a meat-like flavour to soups and stews, followed in the wake of Buddhism—introduced into Japan in the sixth century A.D.— which prohibits the eating of meat and fish. Soy sauce, as we know it, first appeared in the seventeenth century and is now sold all over the world.

Like *miso*, a fermented soybean paste, *shoyu* is made from soybeans, is salty in taste and is used to flavour vegetables, fish and meat. Both are prepared from a base called *koji*, fermented by the moulds *Aspergillus soyae* and *A. oryzae*. For *miso*, the *koji* is prepared from rice, although barley or soybeans are also sometimes used; for *shoyu* the base is crushed, roasted wheat. The salty taste arises from the fact that, following the soaking and pressure-cooking of the soybeans, rice or barley, a considerable amount of salt is added, together with a salt-tolerant yeast, *Saccharomyces rouxii*. The salt is added partly for taste, but also partly to control the growth of bacteria which would otherwise contaminate the end-product with toxins.

With just one more addition, a *Lactobacillus* bacterium, the resulting mash is then allowed to ferment for up to nine months in open concrete bays. As in many other fermentation processes a good deal of heat is released, which can prematurely halt the fermentation process; today, large fans are often used to cool the fermentation down. When fermentation is over, the resulting material is taken to presses, where the dark-brown *shoyu* is squeezed out, pasteurised and bottled. Because of the pasteurisation and high salt content, soy sauce keeps for months without refrigeration.

A Ferment of Shrimps

Fermentation has long been exploited as a means of preserving food. Fish and shrimps, for example, while excellent foods are highly perishable, particularly in a tropical climate. Many are dried and salted, but in Southeast Asia large quantities are also fermented. Again, a high salt content inhibits microbes which would ordinarily cause putrefaction. In contrast to *shoyu*, though, where an injection (or inoculum) of micro-organisms is used to get the process started, the pickling of the fish is triggered by micro-organisms and enzymes found in fish guts and flesh.

Another widely used form of fermentation must also have been easily discovered by accident: leave an uncovered jug of milk standing at the right temperature, and the bacteria naturally present in the milk convert the lactose into lactic acid. This natural lactic-acid fermentation results in a range of stable yoghurt products which resist putrefaction by many food-spoiling or disease-causing micro-organisms.

Yoghurt has long been a basic food in many countries. In Russia, for example, *kefir* grains (consisting of a lactobacillus and a yeast growing symbiotically) are used to trigger the fermentation of milk, producing an acidic, carbonated milk. Sour milks or yoghurts have often been boiled with ground wheat, as in the case of Egyptian *kishk* and Greek *trahana*, with the resulting mash dried in the sun, producing a highly nutritious food which can be stored for years. Modern yoghurt is made by inoculating milk with *Streptococcus thermophilus* and *Lactobacillus bulgaricus*, and incubating the resulting mixture at 45°C (113°F).

Other traditional fermented foods include India's *idli*, an acid, protein-rich steamed bread; *tempeh*, which is produced by fermenting soybeans with *Rhizopus* fungi, and is a wholesome, nutritious food containing about 40 per cent protein and vitamin B_{12}, which is often lacking in vegetarian diets; and the acidic porridges which are a staple food in many parts of Africa.

Far from spoiling such foods, fermentation often actually enhances their nutritional value. For example, the native maize or sorghum beers found in many parts of Africa would be dismissed out of hand by any self-respecting brewer from one of the developed countries, because they tend to be cloudy, sour and gruel-like in consistency. They look like this because they are still actively fermenting, and contain undigested starch, yeasts and other micro-organisms. The fermentation adds such vital trace vitamins as riboflavin and nicotinic acid to the grains, so that people who consume quantities of such *kaffir* beer tend to escape deficiency diseases like pellagra.

Ask most people in the developed world which of the foods they eat are produced by fermentation, and many will be unable to think of any; cheese, yoghurt and bread will probably be mentioned most frequently by those who can. Add in drink, and many will think of beer, cider, wine and spirits. But many other elements of our diet also involve an element of fermentation—coffee, for example.

After oil, coffee is probably the world's most widely traded commodity, and roughly half of the coffee traded has been produced from fermented coffee beans. The 'wet method' of coffee processing involves the pulping of the fresh fruit, followed by a fermentation stage to remove the remaining pulp from the coffee seed. Fermentation is used also in cocoa production, where it is vital to the development of the full cocoa flavour. An enormous amount of heat is released in the process, so that the fermenting pulp has to be turned frequently to ensure a uniform fermentation.

But, even though biotechnology has played a central role in the evolution of our diet, the new genetic-engineering companies have tended to give food and drink targets a wide berth. Even Biogen, which had started off believing that genetic engineering might have something worthwhile to offer to the food and drink industry, ended up with its focus back on pharmaceuticals—losing the backing of Grand Metropolitan in the process (see page 70). Part of the problem, as Grand Met.'s managing director Anthony Good pointed out at the time, is that the profit margins

in the food and drink industry 'do not compare with the pharmaceutical industry. Yet any new product is likely to be treated by
licensing authorities as new and novel, and therefore subject to
the same approval procedure as pharmaceuticals. That makes
their development just too expensive.'
The experience of such companies as G. D. Searle (page 149)
and Ranks Hovis McDougall (page 152) lends support to this
argument, but there are many areas where biotechnology can
help the food industry in the short term. For example, new
bio-tests are being developed to pinpoint food contaminants,
such as *Salmonella*. Whereas some commonly used tests can take
five days to complete, with their interpretation being a matter of
subjective judgement, some of the enzyme immunoassay tests
now being developed give a result in less than 36 hours.

Exploring the Milky Whey

At the same time, the new biotechnologies are also making
significant inroads into some traditional bio-sectors.
If you want to see cheese made in the old way, visit the Basque
country of southwest France. There small-scale cheesemakers
turn out maybe 50 two-kilo (4.4 lb) *ardi gasna* cheeses a year,
making them from raw, unpasteurised ewes' milk. The milk is
curdled with rennet, which contains the enzyme rennin and is
extracted from the intestines of young lambs. Sadly, the bacteria
which give the cheese its distinctive tang can also destroy it: on
average, perhaps one in ten cheeses go bad.
If, instead, you want to see modern processed cheese being
made, you could visit a company like Dairy Crest. There you
might also learn something about the obstacles which such
cheesemakers face in their attempts to harness modern biotechnology to their trade. Cheddar cheese, for example, needs to be
stored for six months if its lactic acid bacteria are to have enough
time to produce that cheddar tang. This is a problem for the
cheesemaker: it locks up capital and in the early 1980s cost about
£140 a ton, about 8 per cent of the resulting cheese's price to the
wholesale trade.
Help seemed to be at hand. 'Researchers at the National
Institute for Research in Dairying (NIRD) have been able to
shorten the ripening period to a couple of months with a commercially available enzyme,' reported Stephanie Yanchinski in *Bio/
Technology*, 'a change that could save the dairy industry £34
million a year. But the new kind of cheese is slightly bitter and
too crumbly.' Further research is therefore needed to find ways of

bypassing these problems, possibly using genetically engineered starter cultures to speed the process.

But there is a cloud on the cheese-making horizon. Enzymes which are used in foods may fall foul of new regulations requiring toxicity testing, greatly increasing the time and expense involved in their development. And it is still far from clear what the reaction of the regulatory authorities will be when genetically engineered foods start to move towards the marketplace.

One possible test case could centre around the genetically engineered rennin produced by Genencor, the joint venture between Genentech and Corning Glass Works. Working closely with Chr. Hansen, a company that claims to be the world's largest supplier of rennin (which it buys from Novo Industri), Genencor completed the world's first large-scale trials of cheeses made with rennin produced by genetically engineered bacteria.

The cheeses, apparently, had an excellent flavour and texture, quite as good as those prepared with commercial rennin. The US biotechnology company Collaborative Research received the first UK patent for recombinant rennin in 1984, which the company claimed was the world's first patent on rennin and also the first patent on an industrial enzyme. The genetically engineered version has extra attractions: it could be available in unlimited quantities—unlike calf rennin (extracted from the stomachs of calves) or lamb rennin—and there would be one less reason for vegetarians to say no to cheese. As far as regulatory clearance goes, Genencor claims that this new rennin is indistinguishable from the natural version.

Meanwhile, another biotechnology company, Bio-Isolates, had been looking into the extraction of proteins from the whey, often treated as a waste material, produced during cheese-making. When curds are separated from the whey, about one-fifth of the milk proteins are typically lost in the discarded whey. Bio-Isolates, whose extraordinary financial history is described on pages 50–52, has a patented protein-extraction process which promises to reclaim this protein.

The basic process, whose product is a protein extract called 'Bipro' or 'Bipro Dairy Albumen', involves passing the whey to an ion exchanger, contained in a stirred reactor, where the protein is adsorbed onto the ion exchanger, leaving the spent whey to be discharged. The protein is then washed off the ion exchanger with clean water, concentrated by ultrafiltration and spray-dried.

The resulting product is a powder which, according to Bio-Isolates chairman Douglas Palmer, 'is probably the best food on earth. It has 70 per cent of the protein in mother's milk, so it comes close to what Nature has decided is right for us.' Although tasteless, Bipro can act as a flavour-enhancer: it can also act as a binder in such foods as hamburgers, and it was originally proposed as a substitute for egg-whites in baking.

The reaction of the baking industry was not all it might have been, however. 'They sent us samples,' said Dr John Randall, chairman of the Avana bakery group. 'It was a perfectly acceptable product, but no better than egg-white, and egg-white was 20 per cent cheaper.' To add to the company's problems, a lower-quality protein sold by the Milk Marketing Board was available, costing about £2,800 a ton as against about £4,000 a ton (then) for Bipro.

Unruffled, Bio-Isolates began to move up-market. 'Bipro is a vastly superior product,' Palmer retorted, 'with unique properties. Judging it on price is a futile exercise. We're now aiming at the health and diet markets. It's going to be a long time before we tackle food processing.' The company has also been looking at possible applications of the process in the recovery of proteins from soybean processing and of blood proteins from slaughterhouse operations.

A full-scale Bipro plant could also generate energy. Bio-Isolates signed an agreement with Dunlop Bio-Processes, under which both companies agreed to collaborate on methane-from-whey technology. To give an idea of the methane potential, Bio-Isolates director George Howett explained that 'if a factory produced 8,000 tonnes a year of cheese, it would produce about 80,000 tonnes a year of whey waste. From this could be extracted about 300 tonnes a year of Bipro, while a volume of methane equivalent to about 600 tonnes of oil a year could be generated from the effluent.' This gas would be produced by the anaerobic digestion of the whey, using bacteria which flourish in the absence of oxygen.

Throughout, Bio-Isolates had been working closely with the University College of Swansea's biochemistry-engineering research group, which later became the nucleus of Biotechnology Centre Wales (BCW). 'We're planning alternative markets for milk products,' explained BCW director Dr Rod Greenshields. BCW, in fact, plans a two-pronged attack. First, it will look at such novel uses of Welsh milk as the production of yoghurt, much of which is currently imported from France. And, second, Greenshields believes that Wales could produce a milk liqueur to rival Bailey's Irish Cream, which holds a dominating position in the UK liqueur market. Meanwhile, other biotechnologists had their sights on different alcoholic targets.

A Fizzy Future

Although the work is still at the top-secret experimental stage, it is known that the Champagne region's official Comité Interprofessionnel de Vin et de Champagne (CIVC) is working with the world's largest champagne producer, Moet et Chandon, on a new, faster champagne-production process. By encapsulating yeasts in a gelatin membrane they hope that, while the yeast capsules will give the champagne its usual bubbles during its second fermentation, the yeasty sediments will be much easier to remove. The whole process should be much faster than the traditional method of *remuage*.

'It is not a revolution,' noted M Yves Bernard, chairman of the champagne makers, 'it is an evolution.' Indeed, wherever vintners, distillers and brewers have held forth on the subject of genetic engineering and the other novel biotechnologies, they have stressed that they are using new technology not to offer new products but to cut costs and maintain quality in their existing products.

At the same time, however, many companies recognise that research-based innovations, as United Breweries president Poul Svanholm put it, are 'the only way we can survive in a very competitive industry'. United Breweries, which won its technological spurs in 1883 when it pioneered the pure yeast strains now used throughout the industry, comprises two of Denmark's leading brewers: Carlsberg and Tuborg. In 1982, they launched Carlsberg Biotechnology to develop new technology and commercialise enzymes and other products of their brewing research. United Breweries researchers are now developing dozens of new yeasts, each tailored to the barley and hops of different regions of the world.

New terms are beginning to surface in the brewing industry, such as 'killer yeasts' and 'rare mating'. A hint of what may be in the pipeline was given by genetic engineer Chandra Panchal of Canada's Labatt Brewing Company. He told a conference in New Delhi that traditional *Saccharomyces cerevisiae* yeasts could be vastly improved if they included genes from a starch-digesting yeast, *Schwanniomyces castelli*, which could help them break down the calorie-adding starches, dextrins, lactose and other such products in the ferment.

The UK Brewing Research Foundation is just one of the organisations now working on the development of diet beers. Its new strain of yeast breaks down dextrin compounds, which are normally left untouched by conventional yeasts, producing a low-carbohydrate, low-calorie brew. The resulting hybrid worked well, but the resulting beer was a gastronomic disaster, tasting of phenol. So the researchers proceeded to identify the gene responsible for the off-flavour and eliminate it. Dr Roy Tubb, who leads the Foundation's genetics-research team, has suggested that a similar approach could improve the ability of yeasts to impart flavour to a brew, or to increase the height of its 'head' of foam.

The team used rare mating, which is an updated version of the brewing industry's traditional yeast-fusion techniques. But, instead of mating two cells to produce a hybrid cell with characteristics derived in a random fashion from the nuclei of both parent cells, rare mating crosses normal brewing strains with mutant strains whose nuclei will not fuse in the normal way. As a result, in a process called 'cytoduction', the hybrid cell can be given genes contained in the surrounding cytoplasm of the mutant strain, without being contaminated by less attractive genes contained in the mutant's nucleus.

The 'killer yeasts' which Panchal and others have been developing are also being worked on by brewers around the world. Some *Saccharomyces* strains naturally act as 'killers', by secreting a toxin, zymocin, which destroys wild yeasts and bacteria that might otherwise contaminate the brew, and genetic engineers are now trying to transfer this ability to some of the more commonly used brewery yeasts. Rare breeding is the preferred approach, since the zymocin gene is found only in the yeast cytoplasm. A key application is likely to be in real ales, where the yeast remains active in the barrel. It remains to be seen whether or not real-ale drinkers will be happy to find genetic engineers fiddling around with their favourite brew.

Meanwhile, many of the more traditional fermentation-based companies have been moving into the new biotechnologies in an attempt to develop new businesses. Some, like Distillers and Whitbread, are underwriting university research in such areas as plant and yeast genetics (see page 193), while others are setting up new commercial joint ventures designed to exploit skills or technologies they already possess.

Allied Breweries, for example, tried to commercialise a continuous-fermentation process, originally developed for beer production, through Alcon Biotechnology—aiming at the fuel-alcohol market (see page 163); and Grand Metropolitan, having dropped out of Biogen, has launched its first new biotechnology venture, Biocatalysts (see page 196). 'Although nothing is going to happen overnight,' as Bass director Dr Tony Portno put it, 'biotechnology is an area we can't afford to ignore.'

In the USA, meanwhile, a fair number of the new start-up

companies have attracted funding from major brewers, as when the Colorado brewers Adolph Coors invested in Synergen, or have linked up with companies in this sector, as Calgene linked with brewing consultants Owades & Co. Coors had also announced that it planned to market a new range of metal-cleaning products, developed jointly with Petroferm USA. These are seen as environmentally safer substitutes for the chlorinated solvents used to process metals for the food- and beverage-container industries.

Back on the brewing front, while some brands of beer have been showing rapid growth, the world market as a whole has tended to be rather sluggish. 'While it is unlikely that brewers will become pharmaceutical manufacturers,' noted Dr Tubb of the Brewing Research Foundation, 'there is nothing to stop them moving into areas related to their existing business.'

Brewers Take to Drugs

Strikingly, however, some brewers *have* moved into the pharmaceutical field, forming joint ventures with genetic-engineering companies. The world's largest brewer, Anheuser-Busch, signed a three-year research agreement with Interferon Sciences in 1983, focusing on the development of yeasts for the production of recombinant interferons. And Japanese brewers have been moving along a parallel course: Kirin, for example, one of Japan's leading brewers, has a joint venture agreement with Amgen covering the development and marketing of the hormone erythropoietin, while Suntory has marketing rights in Japan for Schering-Plough's gamma interferon.

Japan, in fact, is probably the best example of the way that food and drink companies have been exploiting their fermentation skills to break into pharmaceutical markets. An astonishing 80 per cent of food companies list pharmaceuticals as top priority in their biotechnology R&D programmes. Food companies have, moreover, the highest proportion of biotechnologists on their R&D staff and the most comprehensive culture collections.

There is no single key to the food and drink industry's enthusiasm for healthcare. The most obvious explanation is that the industry, with its roots in such fermentation products as *sake* and *shoyu*, has been well placed to grasp the potential of the new biotechnologies. Certainly the fermentation skills of companies like Ajinomoto, Kyowa Hakko and Suntory would place them high on anyone's listing of the world's top biotechnology companies, but other factors have also been at work. The move into biotechnology has been a question of both technological pull and market push, with many food and drink companies only too aware that their traditional markets are highly vulnerable.

Among the stranger new entries is Gunze, a textile and raw silk-spinning company. With a long history in silkworm breeding and in the production of synthetic silkworm feed, Gunze has considerable experience with anti-growth and anti-cholesterol agents, both of which are important in silkworm breeding. It intends to exploit this experience by developing novel human healthcare products.

Even companies like Kikkoman are now moving into the medical field. Yet this transition does not seem particularly strange to the Japanese. Ajinomoto itself points out that its corporate history can be tracked back to a woman who, in the late 1880s, extracted iodine from seaweed at a Pacific coast laboratory, an activity which the company notes would probably now be described as medical biotechnology.

Among the brewing and distilling companies moving into healthcare are Kirin, Sapporo, Suntory and Toyo Jozo. Suntory's biomedical research institute was established in 1979, when the company first began to move into pharmaceuticals, and broke new ground in 1981 with the production of the opioid peptide, alpha-neoendorphin. Suntory is also competing with such companies as Kyowa Hakko, Shionogi, Takeda and Toray-Daiichi in the race to bring gamma interferon to market.

In contrast to some of the mainstream food companies, such alcoholic-beverage producers are moving into biotechnology from a position of strength. Kirin Brewery, for example, is a highly profitable concern with more than 60 per cent of the beer market. Like the second largest brewer, Sapporo, Kirin has built new biotechnology research facilities and is moving into the interferon business. Sapporo is doing basic research on the extraction of anti-cancer agents from rice bran.

A number of dairying companies are also moving aggressively into medical biotechnology, including Snow Brand and Meiji Milk. Snow Brand, in fact, is one of the best examples of a food company making the transition to medical biotechnology by way of health foods and foods with some medical effect. The company's new biotechnology laboratory, at Ishibashi, is among Japan's finest. Snow Brand already sells milk products which are designed for patients suffering from sensitivities to such milk ingredients as phenylalanine and histidine. But it is now also moving into products designed to treat heart disease and cancer, and into plant biotechnology.

Ultimately, and inevitably, there will be a shake-out among

these companies, just as there was in antibiotic production several decades ago. A fair number, however, will develop profitable speciality businesses. S. B. Foods, for example, is a spice company which has spent millions of dollars on a new research facility near Tokyo, planning to look at ways in which biotechnology can help improve spice production.

Things Go Better with Aspartame

'All we are trying now,' said Ajinomoto president Katsuhiro Utada, with what looked like undue modesty, 'is to improve efficiency in the production of amino acids by using biotechnology.' Ajinomoto, in fact, had announced in 1982 that it had used cell-fusion techniques to develop two high-yield hybrid bacteria which, it hoped, would revolutionise the production of two amino acids, lysine and threonine.

Amino acids have always been present in foods: *shoyu*, for example, consists of about 40–50 per cent amino acids, 40–50 per cent peptides, and less than one per cent protein. Now a growing number of US and Japanese companies are wondering how biotechnology might boost amino-acid production.

Japan dominates the world market for amino acids, which are used as additives in animal and human foods. When I visited Kyowa Hakko, for example, it had an astonishing array of bioproducts either in production or on trial, including 20 amino acids, 100 or so antibiotics, about 30 enzymes, a dozen organic acids produced by fermentation and organic synthesis, half a dozen vitamins and growth factors, and about 15 products based on nucleic acids. More recently, too, there has been an increasing use of pharmaceutical-grade amino acids in products designed for use in intensive care units, where patients need intravenous feeding.

The best-selling amino acid is glutamic acid, better known in the form of its sodium salt, monosodium glutamate. Some 300,000 tonnes are produced worldwide each year using a strain of *Corynebacterium*, first isolated by the Japanese in the 1950s. Ajinomoto has been using recombinant-DNA techniques in an attempt to boost the ability of these bacteria to synthesise and excrete glutamic acid. Other key amino acids include aspartic acid, methionine and phenylalanine. Aspartic acid and phenylalanine are the two main ingredients of the new supersweetener, aspartame.

Genex has developed bio-processes for producing both of these amino acids, and is deriving much of its revenue from sales to G. D. Searle, which markets aspartame as Equal (used as a granulated sugar substitute) or NutraSweet (used to sweeten diet versions of Coca-Cola, Pepsi-Cola and Seven-Up). Genex, whose competitors in the production of these amino acids include Ajinomoto and Peirrel, has also come up with an even sweeter-tasting amino acid, serine.

Oddly enough, aspartame is a food additive with roots in medical research. A Searle chemist working on a new ulcer medicine in 1965 did what all school-age chemists are told not to do: he licked his fingers. To his surprise, he found them extraordinarily sweet. Today, after a great deal of intervening R&D, aspartame has been launched by G. D. Searle with a $40 million advertising campaign, the first manufactured food additive to get such star treatment.

Aspartame eliminates the unpleasant after-taste associated with saccharin, producing a sweeter, more appetising drink. No new sweeteners had been cleared for marketing in the UK since the banning of cyclamate in 1969, but in 1983 the Government cleared three: Searle's aspartame, which is about 200 times sweeter than sugar; Hoechst's Acesulfame K, which may not be as sweet as aspartame but is more stable; and Tate & Lyle's Talin, which is 3,000 times sweeter than sugar and comes from a rare African fruit. In a taste of things to come, the gene for this sweetener, thaumatin, has also been cloned into E. coli and into yeast at the Unilever Research Laboratorium in the Netherlands.

Unfortunately, like cyclamates and saccharin, aspartame has run into health concerns, although these do not involve cancer as in the case of the other two. Professor Richard Wurtman of MIT has suggested that phenylalanine could cause problems in those suffering from an inability to metabolise it. This can lead to a severe mental deficiency disease, phenylketonuria. Indeed, Snow Brand markets phenylalanine-free milk for precisely this reason. Wurtman is also worried that consuming large quantities of aspartame could cause depression, sleep problems, headaches and even seizures, since phenylalanine is a precursor to mood- and behaviour-influencing substances in the human body (neurotransmitters). Ironically, aspartame also seems to interrupt production of an appetite-controlling neurotransmitter, so that eating aspartame may actually increase the desire for sweet foods and drink—not exactly what the dieter wants!

Other companies see attractive future markets in the enzymes field. Imperial Biotechnology, for example, sees 'enough crumbs falling from the table' of such major enzyme producers as Gist-Brocades and Novo Industri (see page 195) to support a healthy business. Among the food-related ideas it is pursuing are flavour enhancement, an acceleration in food maturation, and a biological replacement for the mechanical methods currently used to crack open nuts and seeds prior to oil extraction.

Back in the sweetener field, enzymes are vital in converting glucose into fructose, a process first developed commercially in the 1960s. Fructose is used by many soft drink producers, including Coca-Cola and Pepsi-Cola. Three enzymes are used to produce fructose from corn: alpha-amylase, glucoamylase and glucose isomerase. Early on, the final stage of the process, using glucose isomerase, was done in a batch reactor, but in 1972 a continuous system was developed which uses immobilised glucose isomerase. This process is still the world's largest immobilised enzyme process, with a big plant able to convert about 900 tonnes of corn starch into high-fructose corn syrup every day.

The new biotechnology companies have been investing a good deal of time and effort in projects designed to upgrade the performance of the micro-organisms which produce such enzymes; indeed, problems with its fructose contract did much to derail Cetus in 1982 (see page 47). Other companies are working on products like rennet, which has been cloned into bacteria by such industry leaders as Celltech, Collaborative Research, Genencor and Genex. Genencor, as we have seen, is already testing cheese made with recombinant rennet.

Genetic engineers also see potential applications of their trade in the production of a number of vitamins, which are currently either synthesised chemically or isolated from natural sources. It is thought that genetic engineering could help develop more productive microbial routes to such vitamins as B_{12}. Blue-green algae, for example, might be developed as a source of vitamin E, although a great deal of work still remains to be done on the biosynthetic routes by which micro-organisms produce vitamins.

Food flavourings are another likely area of interest for biotechnologists. W. R. Grace has already invested in Synergen, enabling the company to screen large numbers of micro-organisms for possible flavourings and other food additives. And Elf Aquitaine, together with its subsidiary Sanofi, bought the Wisconsin company Dairyland Food Laboratories as a route into the US market for such products. Confronted with Grand Metropolitan's view that genetic engineering is still too costly for the food industry, Sanofi president Dr William Crouch replied: 'When you have an ongoing genetic engineering R&D effort in other fields, the cost versus return would be worth it.'

THE NEW MENUS 151

Menus for the Single Cell

'When people say it's scandalous—a company of your size not in biotechnology—I say we've been there.' So said Professor John Cadogan in 1983, explaining BP's recent reorganisation of its R&D activities to the *Financial Times*. The giant oil company's disastrous early experiences with single-cell protein (SCP)— described in Chapter 10—had not totally extinguished biotechnology in the BP empire. Indeed, Cadogan himself had started a small genetic-engineering group at the company's Sunbury-on-Thames research centre, to serve BP Nutrition, the £500 million-a-year business which grew out of BP's aborted SCP-from-oil project.

The term 'single-cell protein' covers the dried cells of such micro-organisms as algae, bacteria, moulds and yeasts. Archaeologists tell us that the Aztecs harvested the alga *Spirulina* from alkaline ponds, and dried *Spirulina* cells are still eaten by people living around Africa's Lake Chad. There is growing interest in this relatively simple technology: pilot-scale trials have been done in the alkaline waters of Lake Texcoco, and the Israeli company Koor Foods is also working with such algae.

Koor's main current algal project, however, focuses on *Dunaliella*, a green alga which the company grows in large outdoor experimental brine ponds. This alga has adapted to high-salt environments by producing large amounts of glycerol—which happens to be a vital raw material in the production of chemicals, detergents, cosmetics, tobacco, pharmaceuticals and explosives.

SCP has been produced by photosynthetic algae growing in water and by non-photosynthetic bacteria, fungi and yeasts growing on such feedstocks as molasses, methane, methanol, ethanol, cheese whey, cassava starch and a range of agricultural and forestry wastes. Indeed, the fact that some of these micro-organisms could thrive on waste materials encouraged many people to hope that the food problems of the Third World could be solved by SCP technology. So far, at least, this has proved a distant hope: SCP technology is highly capital-intensive, as is illustrated by ICI's 'Pruteen' plant (see page 177).

ICI remains the leading company in this area, and its technology has been attracting interest from the Soviet Union and from oil-producing countries of the Middle East. Despite ICI's best efforts to cut the price of its SCP with improved technology and tailor-made bacteria, it has faced stiff competition from other high-protein animal-feed additives such as soya meal, fish meal and powdered milk. The economics of the process may look better, however, in countries which have both a great deal of surplus methanol on their hands and an agricultural deficit. Both the Soviet Union and the Gulf oil producers fall into this category. The Soviet Union, which is already thought, mainly on the evidence of satellite pictures, to produce about 1.5 million tonnes a year of SCP from paraffin, corn husks and wood waste, has also been looking at the possibility of exploiting cotton wastes. The republic of Uzbekistan alone produces over 10 million tonnes of cotton wastes every year.

Apart from ICI, Phillips Petroleum, the US oil company, is the only other western petrochemical company with a major SCP programme. Unlike ICI, however, Phillips' 'Provesteen' technology exploits a yeast, which it grows on alcohol (potentially based on an enormous range of basic raw materials) or sugars.

Although its early test marketing was aimed at the animal-feed industry, Phillips ultimately aims to break into the market for human foods. The company's biotechnology director, Dr John Norell, said that, while the product's nucleic acid might have to be cut somewhat for human consumption, Provesteen would be highly suitable for human consumption. It will probably be used to fortify flour or rice eaten in Third World countries, he predicted, rather than as a food in its own right.

Give Us Our Daily Fungus

But probably the most revolutionary approach has been that adopted by Ranks Hovis McDougall (RHM), Europe's fourth largest food manufacturer and perhaps best known for such products as Hovis and Mother's Pride. This leading purveyor of the UK's daily bread caught the SCP bug in the 1960s and has spent over £30 million on a fungus which it has learned to knit into passable imitations of fish, chicken and meat. In 1984, it launched a joint venture with ICI, called New Era Foods, to scale up production, hoping to get products onto supermarket shelves by 1986.

More than 3,000 soil samples were collected by RHM, from 1968 onwards, in a world-wide search for a suitable fungus for its project. Sample cultures were flown back to the UK, but RHM finally hit the jackpot in a field a mere six kilometres from its High Wycombe research laboratory.

Originally dubbed A3/5, the micro-organism is *Fusarium graminearum*, a mould. RHM, which produces something like 15 per cent of the UK mushroom crop, pointed out that *F. graminearum* is related to such mushrooms and to truffles, which we eat without a second thought. Apart from being virtually odourless and tasteless, A3/5 contains about 45 per cent protein and 13 per cent fat, a composition which brings it into the same class as grilled beef—and, to RHM's delight in these diet-conscious days, it proved to have a lower fat content than raw beef and to be high in dietary fibre.

A key advantage of mould cells over bacterial cells is that they are typically much larger, which means that they can be extracted from the fermentation broth with ordinary basket centrifuges. Moulds, however, grow much more slowly than bacteria, with a doubling time of 4–6 hours in contrast to the typical bacterial doubling time of just 20 minutes. This has turned out to be a concealed advantage, however, because the slower growth means that the end-product contains less RNA. Whereas RNA levels have reached 25 per cent in some bacteria and up to 15 per cent in yeasts, RHM have succeeded in pushing the nucleic acid content of its new food, for which the generic term is 'mycoprotein', below the acceptable ceiling of one per cent.

In fact, this mould turns out to have an amino-acid content very close to that recommended by the UN Food and Agriculture Organization as 'ideal'. But perhaps the most extraordinary characteristic of this versatile fungus is the way in which it can be turned into a complete spectrum of foods, from soups and fortified drinks through biscuits to convincing replicas of chicken, ham and veal. A lack of suitable flavouring agents initially meant that beef was out of range, and the same problem was encountered in RHM's attempts to produce fish analogues; they sidestepped it by mixing A3/5 with real fish to produce fish cakes and fish fingers.

The key to this adaptability is the fact that the length of the mould fibres can be controlled. The longer the mould is allowed to grow in the fermenter, the longer the fibres and the coarser the texture of the resulting product. The fermenter feedstock is a glucose syrup, as the carbon source, with ammonia to provide nitrogen. The syrup can be produced from any food-grade starch, say potato or wheat, and the process is much more efficient in converting that starch to protein than are farm animals. A3/5 turns each kilo of carbohydrate into a kilo of wet cell mass, which would typically yield 136 grams of protein. A chicken fed the same amount of starch would produce perhaps 240 grams of meat and only 49 grams of protein, with pigs (41 grams of protein) and cattle (14 grams) being even less efficient.

RHM's research director, Dr Jack Edelman, has argued that mycoprotein is an economic way of converting any surplus carbohydrate into foods of a much higher nutritional and commercial value. In the UK the feedstock might be based on wheat, while Ireland could use potatoes and tropical countries might start with cassava, rice or sugar. Many tropical countries, as we have seen, already exploit fungal foods, such as *tempeh*, so that marketing should be easier than in some western countries (although we readily eat such fungal foods as mushrooms and truffles). The main difference between RHM's mycoprotein and *tempeh*, however, is that the former consists entirely of mould cells, while *tempeh* ends up as an amalgam of the soybean substrate and mould. RHM consider this another advantage for mycoprotein, because the slighty mushroomy flavour of fungal protein is felt to be more acceptable than the beany flavour of textured soybean products.

There were three main stages in RHM's development of its new food. Its continuous-fermentation process, which can run for up to six weeks at a time, was the first breakthrough. The second involved the development of new methods for transforming a slurry of mycoprotein filaments into a convincing facsimile of fibrous foods such as meat or poultry. But the third has proved by far the most expensive: all the batteries of tests and trials needed to reassure the regulatory authorities that mycoprotein is safe to eat. Eleven species of animal, including baboons, pigs and rodents, were fed on mycoprotein, in some cases for several generations, to see if there were any signs of toxicity either in the test animals or in their offspring. Human trials were carried out at MIT, with volunteers, and RHM also carried out a testing programme with students and housewives. A3/5's reputation emerged unscathed from all these trials.

The report on all the tests carried out by and for RHM ran to a staggering 26 volumes—two million words. But now that RHM and ICI have linked up to move mycoprotein into large-scale production, the key question is whether the UK market, hardly noted for its alacrity in taking up new foods (with the exception of yoghurt), will take to myco-menus.

The Hi-Tech Grocer

A prediction: by the year 2001 the average grocer will be selling many food and drink products which are unimaginable today. Another prediction: a very considerable number of them will have been produced using some form of biotechnology. And a final prediction: the consumer will be sublimely unaware that biotechnology has anything to do with his or her diet.

Some foods and food additives which have been produced using biotechnology may be identified on the pack, just as NutraSweet's red vortex-like spiral is appearing on many food and drink packs. But, while the food or drink manufacturer may stress that the product is 'natural', it is highly unlikely that genetic engineering or any other form of biotechnology will be mentioned on the back of the pack—or anywhere else, for that matter.

Tomatoes may still come in some form of tin, but the label will tell you nothing about the searches now being mounted in the Andean valleys south of Lima, Peru, for wild strains of tomato which are thought to contain twice as much pulp as the average cultivated tomato. Together with H. J. Heinz, Atlantic Richfield (ARCO) has been trying to develop a tomato containing less water, which would help cut processing costs. And Campbell Soup has been funding work at DNA Plant Technology (DNAP) designed to get 'more cans of soup per dollar'. As David Padwa of Agrigenetics joked, tomato processors ideally 'would like a wooden tomato'.

Anyone who has been hit by one will not be surprised to hear that the average tomato is 95 per cent water and just 5 per cent solids. The less water there is in a tomato, the less useless weight the tomato processor has to pay to have transported around the country. According to DNAP, a tomato engineered to contain just one per cent less water would save soup-makers about $80 million a year!

By using protoplast fusion and full-blown genetic engineering, such companies hope to produce a super-tomato. Campbell Soup, for example, has been growing and studying some strange varieties which could help it in its quest: these include a vigorously growing 12-foot tree which produces white tomatoes, a variety which emits a skunk-like odour which appears to discourage insects and another, from the Galapagos Islands, which is able to tolerate salty soils.

But, even though breeders have advertised them, don't expect your grocery store to stock a 'pomato' next year—or the year after. Plant breeders have certainly mated a potato with a tomato, but undesirable traits can be passed on to the resulting hybrid plant as well as desirable ones. Said Calgene president Norman Goldfarb, 'The chromosomes aren't all that happy. You lose 90 per cent of the yield. You get thickened roots and very small fruits.'

Another possibility: your future grocer's tomatoes, or pomegranates perhaps, may behave differently yet look the same. Now that the gene which controls tomato ripening has been cloned in bacteria, there is a strong possibility that genetic engineers will be able to orchestrate the ripening process to suit the market's needs. Already tomato plants have been bred in which the ripening process has actually been slowed or halted.

In the meantime, the new start-up companies are going to have to come up with products which can arrive in your local grocery a good deal earlier than the turn of the century. DNAP, for example, told investors early in 1984 that it had developed for 'snack packs' new finger-length vegetables tailor-designed to have just the right balance of 'juiciness, crunchiness and sweetness'. Such products, said DNAP's scientific director, Dr William Sharp, could be brought to market in three years or even less.

NINE
The New Fuels

One day, in the not-too-distant future, a microbiologist will duck beneath the clattering rotors of a helicopter on the landing deck of an offshore oil-drilling platform, carrying a box whose living contents could transform the world's oil prospects. Ask one of today's oil-well engineers about the impact of micro-organisms on the oil industry, however, and you are likely to receive a flood of invective: a great deal of effort is currently expended on keeping microbial intruders out of oil wells. But the day is coming when they will be welcomed with open arms.

Modern well-management techniques have dramatically increased oil-recovery rates, with most North Sea reservoirs thought to be yielding about 50 per cent of their oil, compared with rates between 10 and 15 per cent in the first half of the century. But, if you take the world picture as a whole, we are still leaving about two-thirds of the oil in the ground despite the best efforts of the industry.

So-called 'primary' recovery methods, which exploit the natural pressure in an oil reservoir which produces the well known 'gusher', are obviously cheapest—and can extract perhaps 20–25 per cent of the reservoir's oil. But, as you extract the oil, the internal pressure in the underground reservoir drops and the flow slows. At this stage 'secondary' recovery methods may be used, involving the injection of water or gas to boost reservoir pressures and so flush out more oil.

Capillary action tends to draw the injected water into the tiny pores in the oil-bearing formations, rather like ink spreading through blotting-paper, forcing the minuscule droplets of oil out into the main recovery channels. The reinjection of natural gas can also help boost the pressure in the reservoir, and has the additional advantage that the gas can be stored until a pipeline can be built to bring it ashore. Carbon dioxide has also been used, although the expense of the process is indicated by the fact that it can take 112–280 cubic metres (3,950–9,900 cu ft) of gas to flush out a single barrel of oil.

Once the return on water-flushing and similar methods starts to tail off, the time has come to think of 'tertiary' or 'enhanced oil recovery' (EOR) techniques, which are designed to extend the life of the secondary flushing process. One approach involves decreasing the viscosity (stickiness) of the oil by adding chemicals to the water, such as petroleum gases, light hydrocarbons and alcohols. A contrasting approach involves increasing the viscosity of the injected water by adding long-chain polymers to it, enabling it to exert a stronger piston effect on the trapped oil. Among the other techniques exploited by oil-well engineers are: the use of surfactants, which help loosen the bond between the oil and the surrounding rock; the *in situ* combustion of some of the oil, to encourage the rest to flow more readily; and the injection of steam, which has proved one of the most popular techniques. But, even when all these techniques have been used, nearly half of a field's oil may still be underground.

Indeed, Professor Vivian Moses, of London's Queen Mary College (QMC), has estimated that the value of oil which will remain in the North Sea fields will be of the order of £300 billion. Companies have been queueing up to find out how microbial EOR, of which Moses is a proponent, will lead them into the Promised Land of much higher oil-recovery rates.

Injecting New Life into Tired Wells

Although the oil industry currently filters the water it uses for secondary oil recovery, and laces it with biocides to ensure that microbial growth is kept to a minimum (for fear that the microbes will bung up the oil-bearing strata or the injection and oil-extraction wells), the industry may one day inject different micro-organisms into its reservoirs as a matter of routine.

Shell, which was forced to abandon its single cell protein venture in 1977 (see page 176), is one of the companies which have been investing in microbial EOR research, with a view to field-testing at a cost likely to be measured in tens of millions of pounds. 'But just a small improvement can give you quite a big return on your investment,' said Harry Beckers, responsible for co-ordinating Shell's work in this area.

Having identified micro-organisms which can excrete long-chain molecules, or bio-polymers, Shell intends to develop the polymer-piston approach. But, whichever route an oil company took in exploiting *microbial* EOR, the chances are that it would be more attractive than many of the alternatives on cramped offshore production platforms, where space is at a premium. In contrast to most other EOR techniques, which involve a good deal of equipment and other impediments, the microbial route might eventually involve little more than establishing a 'starter' culture on the platform, perhaps from the contents of our microbiologist's suitcase-sized container, and then pumping it down into the reservoir, where it would undergo the required population explosion.

Once in place, the microbes could be used either to manage the flow of oil through the reservoir, by sealing off certain porous areas, thus maintaining the vital pressures needed to drive oil to the well-head, or in actually converting the oil underground into a range of desirable chemical intermediates or end-products. These might include gases such as hydrogen and methane, liquid products such as acids (including acetic acid), solvents (including alcohols and acetone) and surfactants.

'Microbial fermentations,' explained Professor Moses, 'commonly produce alcohol and carbon dioxide. Increased surfactant production is also a well documented phenomenon of bacterial growth on hydrocarbons. And bacteria secrete polymers. Indeed, the bacterium *Xanthomonas campestris* is already used by Merck and Co. to produce xanthan gum, a thickening polymer used in EOR.'

All of these chemicals are already used to flush oil out of the ground, and their *in situ* production could make a great deal of sense. Shallow, exhausted reservoirs in Eastern Europe and the USA have been inoculated with bacteria in a suitable growth medium, typically molasses, and the wells then sealed off while the fermentation proceeded. When uncapped, most of the wells produced an increased flow of oil.

Shallow wells and reservoirs are one thing, however, the deep fields of the North Sea quite another. One problem which will confront microbial EOR there is the high-pressure environment in which the operation will have to take place. Well pressures can exceed 200 atmospheres in the Forties field and 400 atmospheres in the Magnus field. Some bacteria do in fact show better growth at 200 atmospheres than in normal atmospheric conditions—but then pressure is not the only constraint. Temperature is another, with representative temperatures being in the Forties field about 90°C (194°F) and as high as 120°C (248°F) for Magnus. Such temperatures would be instantly fatal to most micro-organisms. Other problems include the high salinities likely to be encountered where sea-water is used in water-flooding and the anaerobic (oxygen-less) conditions in which the micro-organisms will be expected to flourish.

One problem being addressed by Shell and other research facilities is the competition EOR microbes will meet from other microbes which may already have established a foothold underground, such as *Methanobacterium omealnskii*, found in many oil-bearing formations in Russia. Finding a micro-organism which can tolerate the near-freezing temperatures which it will experience as it is pumped down into the reservoir, and then the high temperatures and pressures it will find when it gets there, is clearly a very tall order.

Genetic engineers are now working on hybrid organisms, although some problems may be tackled by using a succession of microbes rather than trying to design all the desirable characteristics into a single strain. Take, for example, the oxygen problem: one microbe might produce the oxygen-rich environment required by the second-wave organisms which will actually do the work of oil extraction and conversion.

The Alcoholic Beetle

If Volkswagen's vision of the future is anything to go by, only half the cars on the roads at the end of the century will be powered by petrol. In the interests of increased energy efficiency, the car of the future will be made largely from plastics and aluminium, with key parts of the engine made from high-performance ceramics. Sadly for those who forecast a bright future for alcohol fuels, however, Volkswagen believes that by the year 2000 only 3–4 per cent of the world's automobiles will run on ethanol produced from biomass. Apart from petrol (gasoline), which the company expects to account for 52 per cent of cars, the other popular future fuels are likely to be methanol made from coal (23 per cent), diesel (15 per cent), and liquefied petroleum gas (6 per cent). Again, if Volkswagen's crystal ball is working properly, only one in every 200 cars will be electrically powered.

Clearly, these forecasts are very much at variance with those produced by enthusiasts of electrically or alcohol-powered cars. They are useful to bear in mind, however, while we take a look at what has been happening in the fuel-alcohol field during recent years.

'Within a few years the state of São Paulo will be just one enormous sugar-cane plantation, straddled by fly-overs,' was

he prospect for
g countries,
ld by the
Sergio
Centre
is bottled
purest form

ture', although
might arrive. In
e East became the
cars ran on a variety
cancer-promoting) ben-
gas and alcohols. These
rol or as fuels in their own
20th century, most engine
r engines which could readily

been commonly used as fuel, either
ol: methanol, the simplest alcohol,
ethane, the simplest hydrocarbon; and
ohol, which corresponds to ethane, the
drocarbon. The 'gasohol' used in the USA in
energy crises of the 1970s is a blend of 10 per
with 90 per cent petrol (gasoline).

ol is likely to be produced from natural gas for many
nstead of by the older route involving the destructive
ation—pyrolysis—of wood. Ethanol, by contrast, is
already being produced commercially from biomass. It can be
derived from three main biomass sources: sugars, starches and
cellulose. Sugars may be produced from sugar cane, molasses
or sweet sorghum, for example; starches come in such forms as
cassava, corn and potatoes; and cellulose can be found in wood
and agricultural wastes.

The basic alcohol-production process has three main steps.
The first involves the reduction of the material to water-soluble
sugars, a step which is clearly unnecessary for sugar-bearing
plants. The second is the fermentation of the sugars to produce
alcohol. And the third involves the distillation, by boiling, of the
resulting liquor, to separate the alcohol from the water. It is the
same process, in fact, that has been used for untold years to
distil spirits, legally or otherwise. In fact it was ironic that, while
a vast effort had been expended in the USA over the years
before, during and after Prohibition to track down and destroy
illegal 'white lightning' stills in areas like Tennessee, here were

those same distillers queueing up for go
essentially the same processes.

As far as Brazil was concerned, it all beg
at least, when President Ernesto Giesel visite
space Technology in 1975, and was impresse
ordered a massive campaign to replace imp
grown alcohol. Having harvested sugar-can
Brazilians' attitude was summed up by one off
send dollars (which we haven't got) to the Ar
send cruzeiros (which we have got) to the farme

Soon, scores of new alcohol distilleries were
producing about 70 litres of alcohol for each ton
ton) of cane processed, a considerable improve
earlier average rate of 12 litres per tonne (about
ton) of cane. Hundreds of thousands of cars, inclu
thousands of the ubiquitous Volkswagen 'beetles'
verted to run on gasohol or pure alcohol. The proble
the sheer number of alcohol cars on the road, cou
Proalcool's success in marketing alcohol overseas and
table production snags, led to shortages on the domestic

In an attempt to bring demand back into line with sup
Brazilian government cut the price differential between
and petrol, so that, when the fact that alcohol cars burn
fuel was taken into account, the real saving to alcohol-car ow
fell to 15 per cent. Whereas in late 1980 eight out of e
ten car buyers had plumped for alcohol, just one year l
alcohol-powered cars accounted for less than 15 per cent of
cars sold in Brazil. Brazil's dream of seeing one million all-alcoh
cars on the road by 1982 remained just that, with the country's
total alcohol-powered fleet having reached perhaps half that
number by the end of 1981. Yet it is easy to overlook what an
achievement even this figure represented. And other countries,
including France and the USA, were joining in.

In France, for example, the *topinambour*, or Jerusalem arti-
choke, became the symbol of the anti-nuclear lobby's response
to the energy crisis, with one slogan exhorting: 'Put a *topinambour*
in your tank.' France's energy planners also foresaw the production of alcohol from such materials as sugar-beet, maize,
lucerne, alfalfa, straw and forest undergrowth, all of which
were seen as potential contributors to the production of petrol
substitutes—or 'carburols'. The aim had been to provide 25–50
per cent of the country's vehicle-fuel needs by 1990, although
the target was later cut sharply as oil prices tumbled.

Back in the USA, all forms of solar energy R&D, the gasohol
programmes included, were hard hit when the Reagan adminis-

tration came into power in 1980. Although some countries, such as France, have continued to build new bio-fuel plants, a number of companies which had developed new fuel-alcohol technologies found that they could not sell them on the world market. One of these was Alcon Biotechnology.

Alcon Biotechnology was a joint venture between John Brown Engineers & Constructors, the company which built ICI's single-cell protein plant (see page 177), and Allied Breweries. John Brown's most significant contribution to the ICI project had been its expertise in 'sterile engineering', a vital consideration if you want to achieve continuous or near-continuous fermentation. Alcon Biotechnology's fuel-alcohol technology, however, uses yeast rather than bacteria, which means that it can survive acid conditions and antibiotic treatments that would bring ICI's Pruteen plant to a grinding halt.

The Alcon process stemmed from John Brown's conviction that it could apply its experience with continuous fermentation in the brewing industry to the production of ethanol. The brewers had become disillusioned with continuous fermentation by the late 1970s, largely because of the difficulties encountered in achieving a consistently drinkable beer. Allied Breweries had invested more research effort in continuous fermentation than any other brewer, but had still failed to produce a reliably palatable ale.

Alcon came up with a continuous-fermentation process which could be housed in a standard shipping container, and this was duly flown around the world to countries which, like the Philippines, have been trying to produce more fuel alcohol to offset growing oil-import bills. But, although the process was the first to win government approval in that country, sales did not follow. As a company spokesman explained, 'due to a combination of political and economic reasons, out of 47 projected orders none were built'. Alcon Biotechnology may well revive if oil prices begin to soar again, but for the meantime the technology has been put 'on the back burner'.

Brazil Could Run on Five Belgiums

Anyone looking for symbols of the problems facing the fuel-alcohol industry would be spoiled for choice. In Brussels, for example, the European Commission proposed in 1984 that the EEC's vast 'wine lake' be turned into fuel alcohol, only to find that the alcohol would cost an average of $2.34 a litre (about $10/ gallon) to produce, compared with 21 cents (95 cents/gallon) for ordinary petrol. It would often be a lot cheaper, they decided, to simply pour the wine down the drain.

And at Kisumu, on the shores of Lake Victoria, Kenya's biggest white elephant quietly continued to rust. Millions of pounds had been spent on the alcohol-from-molasses plant, but it proved too costly and too sophisticated. Still, the news from Kenya was not all bad. The country's first gasohol-plant project, the little known Agrochemical Food Company plant in Muhoroni, came on-stream late in 1982, supplying ethanol-based gasohol to garages in the Nairobi area.

New fuel-alcohol plants have continued to come on-stream. A plant producing 20,000 litres (4,400 gallons) a day of ethanol from whole-wheat opened in Skaraborg, central Sweden, in 1984. The alcohol is being sold to Sweden's OK petrol company and is incorporated in its petrol (4 per cent alcohol to 96 per cent petrol). Interestingly, the dried effluent from the process forms a protein-rich animal fodder, while carbon dioxide and starch are other by-products. This was the first application to a grain feedstock of Biostil, the integrated fermentation and distillation process originally developed by Alfa-Laval and later taken over by AC Biotechnics, a 50:50 joint venture between Alfa-Laval and Cardo. Earlier Biostil plants in Brazil and Australia had been sugar-based, with a further six plants under construction at the time—two in Brazil producing alcohol from cane syrup; one in Pakistan using cane molasses; and plants in France and West Germany being designed to use sugar-beet molasses.

However, even though fuel alcohol may make sense in countries with an energy deficit and a surplus of agricultural products which can be used as an alcohol feedstock, many critics are concerned about the potential conflict between biomass production for energy purposes and for food.

To illustrate the nature of the problem, consider the following. An area the size of Belgium is needed in Brazil to replace a mere 20 per cent of the country's petrol consumption with alcohol. If the USA decided to replace just 10 per cent of its current fossil-fuel use with biomass fuels based on a species like *Euphorbia*, it would need to commandeer an area the size of Arizona for the purpose. And Shell, whose forecasts of the prospects for oil substitutes should perhaps be viewed with a certain degree of caution, has estimated that, if the world's entire crop of maize, sugar-cane, cassava and sweet sorghum were to be converted into alcohol, it would meet a mere 6–7 per cent of the world's energy needs.

And there have been other critics of these early attempts to wean the world from fossil fuel to fuel alcohol. One area of controversy has centred on the question: does a litre of ethanol deliver more energy than was used in its production? Often, it has been shown, the energy balance is actually negative.

If bagasse (the pulp remaining after the sugar has been crushed from sugar-cane) is used as a fuel, or sorghum wastes, with process steam generated by their combustion, the energy balance can look better, but this calculation fails to take into account the energy used in producing and applying the chemical fertilizers and pesticides used in growing such energy crops, and in harvesting them.

Yet biofuels are already extremely significant in some parts of the world, and will become very much more so as our non-renewable energy resources are exhausted. A great deal clearly remains to be done in carefully designing bio-energy systems and in producing more efficient micro-organisms for use in the fermentation process. Some may be able to convert cellulose into alcohols, while others may operate at such high temperatures that fermentation and distillation might even take place simultaneously, saving a good deal of process energy.

Biogen, for example, teamed up with Stone & Webster partly to commercialise a process developed by one of its directors, Dr Daniel Wang, for making ethanol from waste agricultural biomass, including corn stover, using *Clostridia* bacteria. Biogen claimed that it could produce ethanol at perhaps half the cost of other technologies, because the process exploits both the cellulose and hemicellulose in the wastes. But, as oil prices fell, the technology was shelved, although Biogen stressed that it would resume work once oil prices began to rise again.

Other companies were likewise trying to boost fuel–alcohol production, including Mitsui Shipbuilding, which aims to develop yeast strains able to tolerate higher levels of alcohol. Instead of dying when the brew reaches 10–12 per cent alcohol, such improved yeasts might continue to produce alcohol until a concentration of perhaps 17 per cent is reached.

But a number of companies, like Biogen, had begun to question the use of purpose-grown energy crops. If there was going to be competition for grain crops between food and fuel users, they asked, might it not make more sense to base biofuels on wastes? They were soon at work fermenting some very strange materials, from the contents of the normal household dustbin to the horrendous effluents emerging from slaughterhouses.

'Biological Black Boxes'

The Bioenergy process developed by Biomechanics was one of the first anaerobic waste-treatment technologies to come on the market and, while there has been a growing number of competing processes, the company believes that its technology is both more efficient and capable of dealing with a wider range of wastes. Most industrial effluents had previously been treated aerobically, with the bacteria used dependent on oxygen, as in conventional sewage treatment. In the aerobic process, the polluting organic matter is first converted into biomass, in the form of bacterial cells, and then separated from the water in settling tanks. This clearly takes a considerable amount of energy, requires large open tanks, generates large volumes of sludge requiring disposal, and can result in persistent odour problems.

Anaerobic digestion, by contrast, takes place in the absence of oxygen. The organic matter is removed by bacteria which convert almost all of it into a useful gas mixture of methane and carbon dioxide. Whereas in aerobic treatment the resulting biomass sludge often accounts for nearly 80 per cent by weight of the effluent sent into the treatment process, in the Bioenergy process over 93 per cent of the effluent is converted into gas, with only 3 per cent emerging as sludge requiring disposal. The remaining 4 per cent emerges as cleaned water.

The first commercial Bioenergy plant was built in Ashford, Kent, where RHM's Tenstar division produced effluents whose pollution load was equivalent to that produced by a town of 40,000 people. About a year later, a second plant followed, this time in Bordeaux, France. Both plants were designed to treat effluent produced in the processing of wheat into starch, gluten and glucose.

However, these two projects revealed a number of problems with the basic technology, mostly of a biological nature. Within the limited resources then available to it, Biomechanics worked with the University of Newcastle-upon-Tyne to iron out these problems. As part of the continuing development programme, five mobile plants were sited at various industrial locations around the UK. They were used to treat effluents from dairy, cider, pectin, confectionery, yeast, brewing, distilling and chemical plants. The process was used also in Italy to treat effluents from cheese and ham processing, and in Spain in slaughterhouse operations.

Savings on water-authority charges for effluent treatment and energy savings deriving from the use of the methane mean that a Bioenergy plant can show a profit, a very different situation from aerobic treatment, which is typically a total drain on resources. 'This means we can approach firms with a business proposition,' said Biomechanics sales director Paul Ditchfield, 'rather than simply with a process to satisfy statutory requirements.'

However, the biochemical reactions involved in methane fermentations can be extremely sensitive to shocks, such as rapid fluctuations in temperature, acidity or the rate at which the system is fed with new material. This can make it very difficult to obtain a steady flow of biogas from many types of digester. Sometimes, too, most of the energy produced by the digester has been needed simply to keep it warm enough for the fermentation to continue.

In India this problem has often been side-stepped by burying the digester tank underground, thereby insulating it. But of the 100,000 or so biogas plants built in India, most are believed to be no longer functional. Part of the problem has been the taboo about working with human excreta, but it is also true that many of the plants were owned by the richer farmers, who bought up the cattle dung which the poor had traditionally used for fuel. As a result, India's biogas programmes actually left many Indians even worse off than before.

In China, by contrast, the methane picture is rather brighter. Whereas the Indian digesters were often fitted with costly steel caps, the Chinese variety rely on cheap cement, with water pressure used to keep the gas inside. And, while the Indians have had difficulty in persuading villagers to bring their cow dung to the digesters, China's well-organised commune system helped overcome this problem, although it, too, admits that it has had problems. Nonetheless, China reported to the UN in 1981 that it had over 7 million biogas digesters in operation, nearly all of them in its central Sichuan Province. They apparently supply methane to some 30 million people and, in addition, to 150 methane-fired power stations.

Gradually, however, biotechnologists are injecting more science into biofuel production. Another new company which aims to convert wastes into methane is BioTechnica, whose UK subsidiary raised £1.8 million from institutional investors during 1984. 'Our skills are in the microbiological sciences, in dealing with groups of microbes,' explained BioTechnica managing director Dr Stirling Hogarth-Scott. 'We're interested in the use of these skills to resolve environmental problems.'

But, instead of looking for business around industrial-effluent outlets, BioTechnica has an eye for rather more solid opportunities in the waste-disposal field. It has been looking, for example, at ways in which landfill waste-disposal sites can be converted into 'bioreactors' for methane production. 'At the moment,' Dr Hogarth-Scott continued, 'landfill sites are biological black boxes, built by engineers, not by microbiologists interested in improving the rate of decomposition.'

Unlike an area of land planted with energy crops, which can be grown year after year, a landfill site represents a 'biomass mine' that can be exploited only once. Yet the scale of some of these sites beggars the imagination: in the USA, for example, many take in more than 5,000 tonnes of refuse a day. It has been estimated that such sites could satisfy one per cent of total US energy demand. That may not sound like a lot, until you recall how much energy the USA actually uses.

In many landfill sites the decomposing refuse generates methane and other gases, which can seep to the surface, causing explosive concentrations in buildings, killing trees and spreading foul odours to nearby communities. New methods developed by the waste-disposal industry, including higher waste-compaction rates, have affected the water content and temperature of the fill material, sometimes resulting in higher gas yields—and aggravated environmental problems.

As a result, the waste industry has been thinking seriously of designing its landfill sites from scratch as giant bioreactors, with gas production a central objective. One estimate produced by Harwell suggested that the UK industry alone could produce gas equivalent to five billion tonnes of coal.

Overall, an energy source with a relatively small contribution to make, but a contribution which is not to be sniffed at.

Taking a Leaf out of Nature's Book

How do rabbits do it? Renowned for their reproductive abilities, rabbits daily perform a far greater feat which scientists would dearly like to understand better. Every time a rabbit, cow or other herbivore bites off a mouthful of grass, it is performing the first step in a process which converts cellulose into energy. It is estimated that something like a billion tonnes of cellulose go to waste in the USA each year. What can rabbits teach us about exploiting this abundant, tricky natural resource?

The key to the rabbit's ability to break down cellulose is an enzyme, cellulase, found in bacteria which inhabit the animal's intestines. The problem is that these bacteria operate too slowly to be of much interest to those who want to develop commercial fuels; but a team of scientists at Cornell University have shown how genetic engineering can be used to boost the efficiency of the process. They isolated a gene which codes for cellulase in a bacterium called *Thermomonospora YX*. Following on the heels of so many other genetic engineers, they then inserted the gene into *E. coli*, which promptly started to churn out quantities of cellulase. By further tinkering, Professor David Wilson and his colleagues managed to boost *E. coli*'s cellulase production about 50-fold. Although this work still has a considerable way to go before oil magnates start having sleepless nights, it does show how recombinant-DNA technology may eventually transform biofuel production.

The energy crises of the 1970s sent biotechnologists racing to their laboratories, from which, a short time later, a flood of fascinating ideas began to pour. Most involved covering considerable areas of land with new types of solar-conversion devices. An algal bioreactor proposed by Professor John Pirt, which led some scientists to argue that Pirt appeared to be rewriting the laws of photosynthesis, involved a tubular, transparent reactor containing a continuous culture of *Chlorella*. This was fed with ammonia, salts and pure carbon dioxide. The resulting algal biomass, Pirt suggested, would have a high starch content and could be converted into liquid fuels and oxygen. Using the reactor, Pirt estimated, energy equivalent to 100 tonnes of coal could be produced each year in the form of algal biomass from just a single hectare (2½ acres) of land. Sceptics argued that a more accurate figure might be 30–50 tonnes, but Pirt went on unabashed to predict that 100 million tonnes of coal equivalent —or 30 per cent of the UK's fossil-fuel requirements—could be produced on a land area of 10,000 square kilometres (3,900 sq miles).

The main focus of such 'photobiological' investigation has been on hydrogen production by whole micro-organisms. For example, when I visited the Colorado-based Solar Energy Research Institute in 1981, shortly before its R&D programmes were devastated by the Reagan administration, Dr Paul Weaver had just returned from a trip to a number of hot springs. He had been looking for microbial talent in some extraordinarily hostile environments. Among the microbes he had tracked down, some were found to be producing hydrogen, which, ultimately, is likely to be an extremely important fuel.

Among the organisms which use the enzyme hydrogenase to produce hydrogen are photosynthetic bacteria, cyanobacteria, green algae, red algae, brown algae, and a number of non-photosynthetic bacteria. In short, a wealth of microbial talent appears to be available if we look hard enough—and in the right places. If, as Professor David Hall of King's College, London, has argued, a photobiological system could operate at an overall efficiency of 10 per cent,

> then the world's total current energy needs could be met by turning over only half a million square kilometres [about 200,000 sq mi] to solar-energy collection, just 0.1 per cent of the Earth's surface. This area is about the same as that of Morocco, Thailand or France, or twice the area of the United Kingdom, or one-fifteenth that of Australia or Brazil, or three-quarters of Texas. It wouldn't be necessary to take over agricultural land, and the sea could provide the water.

In the UK, he suggested, with the total incoming solar energy each year equivalent to 100 times the country's total primary energy consumption,

> we estimate that a tenth of the UK's land area would meet our total energy requirements. In other words, 2 per cent of the land would yield hydrogen equivalent to the country's natural-gas consumption, while 4 per cent would match our petroleum or coal needs. The UK isn't likely to set aside such large areas for its energy needs, but the area of land needed to substitute our natural-gas requirements is interesting.

In the longer term, some photobiologists believe that they could dispense with the whole-organism approach, dispensing simultaneously with the problems associated with keeping such vast collections of micro-organisms alive. Instead, they believe, we could use carefully selected components of suitable organisms.

Take *Halobacterium halobium*, a micro-organism which flourishes in the Dead Sea and which has been the subject of research projects in the UK, Holland, Hungary, West Germany, the USA and the USSR. The main focus of attention has been on a purple-patched membrane found inside the micro-organism. This contains a pigment, bacteriorhodopsin, which harnesses sunlight to pump protons from one side of the membrane to the other. This pumping action also appears to take place even when the cells themselves have been destroyed and only the purple membranes remain. This fact suggests that such membranes might form the heart of low-efficiency, low-cost solar energy collectors for the Third World.

An even more speculative application of the purple membranes would be to produce hydrogen and oxygen *via* photolysis. If plant chloroplasts can do it, the argument runs, surely systems incorporating such membranes could do so too? But many photobiologists are far from convinced, believing, like David Hall, that the best approach may not be to use biological systems direct, but to study them to see how they operate and then to imitate them.

There is clearly no shortage of ideas about ways in which biotechnology might help us make the inevitable transition from non-renewable to renewable sources of energy, but almost none are likely to be commercial in this century. The planet's total reserves of oil, coal and non-traditional sources of oil, such as oil shale and tar sands, are so enormous that the economics of most biofuels (other than those which, like fuelwood, are already widely used) are going to look decidedly unattractive in most parts of the world for some time to come.

It can be very exciting to get a bug to produce hydrogen in a test-tube, or to culture algae in a laboratory reactor, but biofuel enthusiasts face the same basic challenge as all other biotechnologists: eventually you have to scale up your process to something which makes commercial sense.

The experiences of such companies as Cetus and ICI, in fields as disparate as heavy chemistry and single-cell protein production, raise a series of question-marks over many of the forecasts which have been made by the world's biotechnologists in their efforts to persuade investors to fuel their pet R&D programmes.

TEN

The New Production Lines

Anyone who has ever tried to brew beer in an airing cupboard already has a fairly good idea of some of the problems which can plague the commercial biotechnologist—and of the key elements of some of tomorrow's production lines. To produce a palatable beer you not only have to pick the right ingredients and the right yeast, but the fermentation has to be carried out in the right sort of container, kept at a suitable temperature and protected against such sources of contamination as the vinegar fly. Compared with the problems of keeping one of today's continuous-fermentation processes sterile, however, the home brewer's difficulties are as nothing.

If a new bioprocess is to show a profit, fermentation and separation technologists will have had to hit a number of difficult targets simultaneously. Assuming they have got the right micro-organism, with any genetically engineered abilities proving stable in normal fermentation conditions, they will have had to find a suitable feedstock at the right price; to design and build a fermentation system which, while perhaps not flawless, has no significant defects; to work out a suitable fermentation time to maximise their product and minimise spoilage; and, among many other successes, to have worked out cost-effective ways of isolating and purifying the product. A series of tall orders.

Most industrial fermentations, like the home brewer's bucket of beer, are still *batch* fermentations. But, while there may be similarities between the brewer's bucket and the giant fermenters used to produce such biotechnology products as antibiotics or single-cell protein, they are outweighed by a very considerable number of differences. Although companies like Kikkoman still use solid fermentations, almost all the major industrial fermentations today are liquid. So, in a typical batch fermentation, a micro-organism is grown for a number of days in a liquid medium which contains the various nutrients it needs. As soon as a sufficient concentration of product has built up in the microbial cells or fermentation broth, the fiddly, expensive process of extracting and purifying the product begins.

When a batch fermentation is carried out in a bucket or in a small laboratory fermenter, it may simply look after itself, with no further help from the brewer or biotechnologist. Often, however, the fermentation flask needs to be gently shaken, to encourage the aeration of the broth and thereby the growth of the yeasts in it. At the laboratory level of operation, the ratio between the volume of the fermenting broth and the surface it presents to the air at the top of the fermenter permits adequate natural aeration of the broth. It also permits the heat of fermentation to escape, while sterilisation of the whole fermentation system can take place frequently and with relative ease.

The 'scale-up' problems which many of the new genetic engineering companies have encountered in their attempts to produce commercial quantities of particular products stem in large part from the very different conditions to be found in the heart of a large-scale fermentation—which may be anything from 2,000 to 20,000 times the size of the ten-litre (2¼-gallon) volumes found in the standard small laboratory fermenter. A totally new approach may be forced on the biotechnologist.

There are two dramatic differences between this large-scale fermentation and the tests that were carried out at the laboratory bench. First, the ratio between the fermentation's volume and surface area has changed such that the heat generated by the fermentation is now bottled up within the fermentation, unable to escape at a sufficient rate. So the fermentation begins to heat up, often rapidly, with the result that our carefully engineered micro-organisms lose their commercially desirable attributes and, if the overheating continues, die. Second, the rate of oxygen-transfer is much lower, which can inhibit the growth of the micro-organisms we are trying to culture and possibly, by forming anaerobic pockets in the broth, may promote the growth of micro-organisms which can wreck the fermentation.

Simultaneously, we will probably spot a number of related problems. Without some mechanism to stir the fermentation, the broth will become locally depleted of nutrients or contaminated by toxins and other microbial by-products. These may depress or halt further microbial activity. This ultimately abortive fermentation will have taught us a number of unforgettable lessons about commercial biotechnology.

We will have recognised, perhaps for the first time, that we are going to have to spend a great deal of money if future fermentations are to proceed at a suitable rate and produce the product we want. Our fermenter will almost certainly need to be completely redesigned. For a start, it will need support systems for mixing, aeration and refrigeration, and it will also need monitoring equipment to track what is going on in the depths of the fermentation. And there is no point in knowing what is going on if we cannot influence the fermentation conditions: we shall need control systems to steer the fermentation out of trouble.

Sterile Pursuits

One of the first things the successful home brewer learns to do is to sterilise all the tubes, containers and bottles used. If this is not done, the resulting brew may be vinegary or suffer from one of many off-flavours: the first batch of orange wine I ever made ended up tasting of mothballs. Sterilisation is important in most industrial batch fermentations, too, but it becomes absolutely critical if you want to set up a *continuous*-fermentation process, as RHM has done with its mycoprotein technology (see page 152) and ICI with its Pruteen single-cell protein technology. Others, as we shall see, have been less successful.

Tate & Lyle, best known for sugars and syrups, spent something like £20 million during the 1970s on research designed to add value to carbohydrate feedstocks. 'There are those who believe in research, those who don't, and those who do—but only just,' said Professor Chuck Vlitos, summing up the problems he had encountered in persuading Tate & Lyle to set up a new, centralised R&D facility. He had argued the case for using sugar as an alternative to oil feedstocks, pointing out that sugar chemistry was surprisingly flexible. Tate & Lyle accepted his arguments, and the Philip Lyle Memorial Research Laboratory was opened alongside the University of Reading in 1972. Early developments, even before the 1973 OPEC oil shock, focused on new sucrose-based materials such as surfactants (the basis of detergents), intermediates for making resin polymers, agricultural chemicals, pharmaceuticals and food additives.

The initial vehicle for this work was Talres Development ('Talres' being a contraction of 'Tate & Lyle research'). Refining was seen as the critical stage in adding value, so the company built what has been described as the world's first 'biorefinery' at Knowsley, near Liverpool. Vlitos envisaged a range of 'performance chemicals' emerging from the £15-million biorefinery. Tate & Lyle saw a number of potential advantages for its proposed product line, which included thickeners, preservatives, food additives, lubricants and fire-retardants. These included the fact

that such products would be based on cheap, renewable feedstocks, while it was expected that there would be far fewer
toxicity problems than with oil-based feedstocks.

The first product to emerge from the biorefinery was a powerful surfactant made from sugar and vegetable oil. But the undoubted 'flagship' of the Tate & Lyle biorefinery was Biospecialities, a £6-million joint venture between Tate & Lyle and Hercules
Powder, the US chemical company. Biospecialities was established to produce xanthan gum from sugar with a continuous-
fermentation process developed by Tate & Lyle. Gums such as
the alginates and xanthans are used extensively in the paper,
food and pharmaceutical industries and were extracted, until
relatively recently, from natural but ultimately limited sources
such as seaweed. Microbial fermentations can produce both
known gums and, potentially, novel materials. It was Tate &
Lyle's misfortune that it embarked on its biorefinery when the
biotechnology industry was still relatively ignorant in some important areas.

Sadly, sterile engineering, which aims to exclude all sources
of potential contamination from a fermentation system, had not
been given sufficient weight. Following a series of contamination
problems, an engineering study suggested that it would cost a
further £1 million to ensure sterility. This was the kiss of death
so far as continuous fermentation was concerned.

Ultimately, the biorefinery failed to produce any products,
and Biospecialities was forced to sell its plant to the xanthan-gum
market leaders, Merck. They stripped out the drying equipment,
replacing it with their own, and switched to batch production to
cut the contamination problems.

Toprina and the Italian Mouse

It should have been the production line of the future. It should
have turned a low value by-product of oil refining into animal
feed at a time when worldwide food shortages were apparently
looming. It should have generated a booming export market
for single-cell protein plants and for various elements of the
supporting technology. Sadly, as events turned out, BP's single-
cell protein technology failed on all of these counts.

Yet, to be fair, the original idea had a great deal going for it.
BP had worked out two different approaches for feeding yeasts
on oil by-products, with the resulting yeast cells forming the rich
protein concentrate which BP dubbed 'Toprina'. The first of
these continuous-fermentation technologies, developed by a BP

laboratory in France, exploited the paraffins found in gas oil. The original idea was to extract from oil the waxes which otherwise lower the oil's value on fuel markets. The yeasts used were then recognised as a potential source of single-cell protein (SCP). A 17,000-tonnes per year plant was built near BP's Cap Lavera refinery, with the processed oil fed back into the refinery. The second approach, developed by BP at Grangemouth, Scotland, exploited n-paraffins, which are mineral oils used in both food and medical applications.

So enormous was BP's investment in SCP that other major companies were attracted into the field, including Exxon, Hoechst, ICI and Shell. BP, meanwhile, had begun talks with ANIC, part of the Italian state oil group ENI—which had a great deal of Libyan oil on its hands which needed de-waxing because it was rich in n-paraffins. Eventually a 100,000-tonnes per year Toprina plant was built alongside an ANIC refinery at Sarroch, in Sardinia. Foster Wheeler Italiana engineered the plant for Italproteine, a joint venture between ANIC and BP launched in 1971. But then the real problems started.

The new plant, commissioned in 1976, was capable of producing protein equivalent to the output of 750,000 acres of wheat. Toprina had been cleared, following extensive toxicity trials, by the Italian Ministries of Agriculture, Health and Industry. And BP expected no further problems because, while later Italian tests were to show that animals fed on Toprina had up to 71 parts per million (ppm) of n-paraffin in their tissues, WHO was happy to see 100ppm—and the material was already widely used in food preparation. The Italian authorities permitted up to 1,400ppm in rice, for example, while the French permitted up to 20,000ppm in coffee.

And then the Italian toxicologists began to agitate. One scientist claimed to have found a previously unrecorded residue of n-paraffin in the fat of pigs fed on Toprina. At the time, the Italians were keen to develop their own toxicity-testing programmes, a trend summed up by one US toxicologist as 'an unwillingness to accept results unless demonstrated on an Italian mouse'. BP later claimed that the entire problem was due to the fact that it had not funded toxicity tests in Italy.

The battle raged to and fro, with an increasing conviction among BP's negotiators that there was a political, rather than a scientific, attack being made on the whole food-from-oil concept. Meanwhile, there had been growing worldwide interest in BP's technology, with schemes proposed in Saudi Arabia, the Soviet Union and Venezuela hanging on the outcome of the Sardinian project. By 1976, however, the price of oil had risen enormously, while the price of soybeans had not soared to the extent that BP had predicted. Worse, as far as the economics of Toprina were concerned, new breakthroughs in plant science had improved the soybean beyond recognition, making it a much hardier crop plant.

Confronted with all these problems, BP decided to abandon Toprina in 1977. Such was the emotional shock of this reverse, one BP director told David Fishlock of the *Financial Times*, that the subject of biotechnology could not be raised at board level for a considerable time afterwards.

Nevertheless, David Llewelyn, who had been involved in BP's single-cell protein work since 1963, persuaded the group's board that, with the new worldwide interest in biotechnology, it should market the information it had collected during the entire programme. Some buyers, it was felt, would purchase the information simply to find out what not to do, in an attempt to avoid BP's mistakes. Llewelyn devoted a year to producing a four-volume information system on Toprina, distilled from 400 reports and technical documents. This was first offered for sale in 1981, for £100,000 a time, or one-thousandth of BP's own total £100 million investment in Toprina. Despite some interest, no copies had been sold by the time of writing.

A £100-million Entry Fee

A rugby pitch in County Durham was where ICI scientists finally struck gold, isolating the bacterium *Methylophilus methylotrophus* from soil samples. They had spent 13 years hunting for a microbe which would grow rapidly on readily available petrochemical feedstocks, yet would produce a protein concentrate suitable for feeding to chickens, cattle and fish: 'Pruteen'. In total, ICI screened some 10,000 micro-organisms—over three times as many as RHM (see page 152), which had also started its quest in 1968. Some had been collected from North Africa, where they haunted the sand beneath oil-drilling rigs, others from Australian swamps.

Early on, ICI focused on methane as a potential feedstock, largely because its agricultural division had access to plentiful methane from the North Sea. At first sight, this seemed to offer a highly elegant biochemical route from the simplest of organic molecules to the much greater complexity found in proteins. Very soon, however, ICI began to change its mind. Not only were there potential explosion hazards to be coped with when using methane in aerobic fermentation processes, but, more importantly, there were also considerable problems in ensuring that methane, which has a low solubility in water, was equally distributed through the fermentation broth.

The company was already a world leader in methanol technology, however. Using methanol, a partially oxidised form of methane, can cut oxygen demand and heat-generation in fermentation processes. ICI judged that methanol, which *is* water-soluble, offered a much better prospect, if an appropriate micro-organism could be found. The microbe it had found on that rugby field proved to be stable, capable of high carbon conversion when grown on methanol, and free from toxic side-effects.

The nature of the beast convinced ICI that not only would it be better to culture the bacterium on its own, rather than in a mixed fermentation, but that a continuous-fermentation process would be desirable. The implication was that the plant would have to be engineered for a quite remarkable standard of sterility, to prevent any other micro-organisms finding their way into the fermentation. Other SCP producers had picked yeasts precisely because they tend to be much less demanding as far as sterility is concerned.

John Brown Engineers & Constructors was commissioned to build at ICI's giant Billingham site a plant capable of producing 50–70,000 tonnes of SCP a year. The resulting plant, representing the largest sterile fermentation process in the world, works broadly as follows. Sterilised air, water, methanol, ammonia and other nutrients are continuously fed into the single tall fermentation vessel, which is split into 'riser' and 'downcomer' columns, the former bearing a rising fermentation stream, the latter closing the cycle, allowing the fermentation medium to circulate back to the foot of the riser. Air is injected into the lower part of the riser, causing vigorous mixing and, by reducing the nutrient medium's density, promoting circulation throughout the system. The medium releases carbon dioxide and unused air at the top of the fermenter, before moving back down the downcomer. The fermenter design ensures a high level of dissolved oxygen and very thorough mixing.

The fermenter is brought into action by first sterilising the empty vessel with steam, after which it is filled with sterile medium and inoculated with just a few litres of bacterial culture. Within two days, the micro-organisms grow and multiply to the extent that the fermenter contains 60 tonnes of live culture. At this point continuous culture can start, with the flow of nutrients into the fermenter being carefully balanced by the withdrawal of culture medium for cell harvesting. The methanol is injected into the fermenter through about 3,000 separate ports. The entire system has surprisingly few moving parts, to cut both contamination risks and energy costs.

An early problem, however, was foaming, with 1.8m (6ft) of foam sometimes produced by the 'airlift' fermenter. Most commercially available anti-foaming agents would have been broken down during the sterilisation processes which are such a key element in the plant's operation, but scientists working elsewhere in the ICI group had come up with a suitable agent, originally designed for the brewing industry. Once under way, there was no reason, other than low market demand for Pruteen, why the fermentation should not run for as long as a year.

The cell-separation stage also takes place in sterile conditions, so the spent medium can be returned to the fermenter. Continuous centrifuges produce a thick cream, containing more than 20 per cent of cell solids, which is sent to the final drying stage. This produces a granular product, basically brown clumps of the dried bacterial cells, which can either be used direct in feed formulation or ground down to produce a finer powder, for use in the (higher-value) form of a milk substitute.

When the Pruteen plant was first given the go-ahead in 1976, the ICI board expected that it would show a modest return on the investment, but excellent soybean harvests and escalating energy prices have meant that the plant has not been profitable. So ICI has been working on new strains of the basic microbe, developed using both recombinant-DNA and classical genetic techniques, claiming that this was 'probably the first commercial application of genetic engineering on an industrial scale'.

ICI has produced at least one super-strain, by introducing a gene coding for an enzyme able to metabolise methanol more efficiently. This improved microbe could boost the methanol-conversion efficiency of any future plants by 5–7 per cent. Such a microbe would have to go through the same battery of toxicity tests as the original microbe, but ICI has said that, even if the additional yield were of the order of only 3–5 per cent, the extra toxicology work would be worth doing. Future plants could also use oxygen rather than air, which would cut out the electricity bills resulting from the need to compress the air injected into the current Pruteen plant.

If the plant is run at less than full capacity, the cells may make and store carbohydrates, which can result in discoloration of the final product. So ICI, because of the constraints on the markets for the material, has tended to run the plant at full capacity for relatively short periods, stockpiling the Pruteen, and closing the plant down until a new production run is needed. ICI admits that the world of the 1980s has turned out very different from that projected in the late 1960s. 'Most people then were predicting an era of cheap oil and gas, coupled with a long period of protein shortages,' explained Rob Margretts, closely involved in the running of the Pruteen project.

Far from being chastened by the economics of the Pruteen plant, however, ICI now sees the £100 million or so it estimates it had spent on the plant by 1982 (at 1982 prices) as part of its entry fee to the biotechnology field. Certainly it has had more success in selling its SCP technology than BP had: it sold the technology to the Soviet Union in the late 1970s for two giant SCP plants, for example, and hopes for further sales to both the Soviet Union and Saudi Arabia.

ICI has also been looking for other products which can be produced using its Pruteen technology. The first major product it announced was a new biodegradable plastic produced by the bacterium *Alcaligenes eutrophus*. Such bacteria produce a range of polyesters if they are given a feedstock which is deficient in nitrogen. Called polyhydroxybutyrate (PHB), ICI's microbially produced plastic can be moulded, reinforced with inorganic fillers, spun into a fibre or formed into a film. It also has the property of 'piezoelectricity', which means that it undergoes changes in its molecular structure when exposed to an electrical field. Its biodegradability is more likely to prove attractive in the short term, however. Because it is non-toxic, is compatible with living tissue, and is broken down naturally in the body over time, PHB seems an attractive material for making medical sutures and pins, used for reassembling patients after surgery. Another product likely to be made in the Pruteen plant is RHM's myco-protein (see page 152), following an agreement between the two companies in late 1984.

Meanwhile, other companies working in the SCP field, like Hoechst and Phillips, have adopted a different approach to ICI's, aiming to supply protein for human consumption. To satisfy the appropriate regulations, however, the resulting protein would have to have some of its nucleic acids extracted. Hoechst, rather than treating such nucleic acids as wastes, has been looking at ways of turning them into flavourings and flavour enhancers, high-value products which could make SCP economics look rather more appetising.

Calf Stomachs and Bone Char

Producing speciality chemicals as a by-product of SCP production is one way to run a business: another is to aim for speciality chemicals from the word go. Biotechnologists use bioreactors for this purpose, some of them even more sophisticated than the Pruteen bioreactor. But one of the first bioreactors ever used was very simple. When the ancients used calf stomachs to store milk, they found that something inside the stomachs (in fact, the enzyme rennin) turned the milk into cheese. Since then, bioreactor design has advanced considerably, although the basic idea is still to bring some material into contact with enzymes and to promote its conversion into a more valuable product. Remember Tate & Lyle's ill-fated interest in 'biospecialities'? The company is still producing high-value bio-products, most of them using immobilised enzymes.

According to Dr Christopher Bucke, who runs Tate & Lyle's biotechnology programme, the 1970s should have been called 'the decade of the immobilised enzyme'. He has pointed out that two million tonnes of high fructose syrup were produced in the USA by 1981, despite the fact that the product was unknown in 1970. Most of today's high-fructose syrup is made with Novo's whole-cell preparation of *Bacillus coagulans*.

Tate & Lyle has carried out pioneering work on the immobilisation of enzymes, working in close collaboration with the Harwell Laboratory—set up to carry out nuclear R&D, but a prime example of how originally single-mission research facilities have diversified into biotechnology. The problem, they found, was that if they simply stirred enzymes into a process they generally lost them when the process was finished, or were faced with high enzyme-recapture costs. Immobilised enzymes, by contrast, cut reaction times, reduce the formation of unwanted by-products, and simplify enzyme recovery and regeneration.

The basic technique was developed by a Japanese drug company, Tanabe Seiyaku, in 1969. Many different methods have been developed for immobilising enzymes, but some have proved to be unduly expensive, at least as far as Tate & Lyle's sugar chemistry is concerned. The main element in this cost has proved to be the need to regenerate the support material after use.

Tate & Lyle decided that a new approach was needed. Sugar chemistry, it felt, needs a support material which is cheap enough to be thrown away after a single use. It found the ideal material in bone char, already used in the sugar industry. More recently, however, the company has also been using alternative support materials as it begins to move into the rather more difficult field of immobilised cells. Some reactions need more than one enzyme, in addition to the 'co-factors' the enzymes need to do their work. Unfortunately, some of these enzymes are likely to need different conditions to perform properly, so any such mixed system turns out to be a less than perfect compromise. An alternative, illustrated by Novo's whole-cell approach to fructose fermentations, is the immobilisation of complete microbial cells.

One pilot-scale process reported by Dr Peter Cheetham of Tate & Lyle produced a fascinating sugar. This was a continuous-fermentation method, using columns of *Erwinia rhapontici* cells trapped in alginate gel pellets. Fed with concentrated pure sucrose, the entrapped cells, which normally cause crown-rot disease in rhubarb, convert the glucose into isomaltulose. This is a particularly interesting new sugar because the evidence suggests that it may help prevent tooth decay—perhaps because the bacteria that cause dental caries cannot use isomaltulose as an energy source.

Far simpler fermentation processes can also work much more efficiently if the cells doing the work are supported. To take just one example, the Captor pollution-control process developed by the University of Manchester Institute of Science and Technology (UMIST) uses polyester-foam support particles to support the cells which break the effluents down. After the particles are filled with the resulting microbial biomass, they are extracted from the reactor, squeezed like sponges to remove much of the biomass, and returned to the reactor. UMIST has also been looking at the use of this system for culturing plant cells, with initial work focusing on the production of capsaicin from cells of *Capsicum frutescens*.

Fishing in Soggy Porridge

Products such as SCP and plant cells are fairly easy to extract from the fermentation broth, but many of the much smaller products which medical biotechnologists are interested in are proving more difficult to pin down. Sometimes the biotechnologist is looking for the biochemical equivalent of a needle in a haystack.

If even a fraction of the market projections brandished by the new biotechnology start-ups are to be realised, a number of significant 'downstream' bottlenecks will need to be removed— including those found in product extraction, separation and purification. Companies like Sweden's Pharmacia Fine Chemicals see part of their mission, as Pharmacia has put it, as 'helping the biotechnology industry to purify and bring to the public the new products created by genetic engineering'.

Easier said than done. 'The genetic engineers will continue to give us micro-organisms designed to produce specific products,' explained John Curling, general manager of Pharmacia's large-scale chromatography unit, 'and the fermentation engineers will give us bulk products, but in a highly impure form. Now it is the protein biochemists and downstream-processing engineers who will turn fermentation products into pharmaceuticals.' These are the people who go fishing in the broths and sludges produced by fermentation for such end-products as the interferons. For, as Genentech has put it, the 'laboratory creation of a new product is the first step. It only begins to count when the product is purified and packaged.' But this is tricky fishing. As Duncan Low of Pharmacia put it, 'you are faced with getting out the product you want from something that has the consistency of soggy porridge'.

Downstream processing can be divided into three main stages: extraction, separation and purification. Starting from a microbial culture in which the product has been produced either inside or outside the cell ('intra-cellular' or 'extra-cellular' production), it is easiest to deal with extra-cellular products. Intra-cellular products are harder to get at, since, once you have extracted the cells, you still have to digest the cell wall with chemicals or enzymes—or you may choose to mash them up mechanically. Either way, the proteins you are looking for may well prove to have locked tightly onto some of the cellular debris you have created, and some of them may have been denatured (de-activated) by enzymes liberated in the process.

This is not the place for a blow-by-blow account of all the extraction, separation and purification options which are available, but a quick review of some of the techniques used by the Swedish company Pharmacia will give the flavour of what can be involved. The obvious method of extracting cells is to filter them out of the fermentation broth which, in one way or another, is what is done when centrifugation, filtration or ultrafiltration is used. The difference with centrifugation is that the broth is placed in a rapidly spinning container from which the spent broth can escape through holes of an appropriate size, with the cells remaining trapped inside. But let's look at two rather more sophisticated separation and purification techniques: chromatography and electrophoresis.

Pharmacia has developed a number of chromatography options, five of which are widely used. These five techniques are gel filtration (which exploits the size of a molecule to achieve purification), ion-exchange chromatography (which exploits a molecule's electric charge), affinity chromatography (which exploits its structure), hydrophobic chromatography (which exploits its polarity) and chromatofocusing (which exploits its isoelectric point). Pharmacia's range also includes high-pressure liquid chromatography and, one of its most recently developed techniques, fast-protein liquid chromatography (FPLC).

As the experience of P & S Biochemicals shows (see page 190), FPLC can be used to purify many of the key tools and products of biotechnology, including restriction enzymes, hormones and monoclonal antibodies—all of which are highly potent in extremely small quantities and need to be kept virtually contamination-free. This technique cuts the time required for separating such bio-molecules from hours, or even days, to minutes. But some forms of electrophoresis may prove even more powerful, especially if they take place where our planet's gravitational field is beginning to weaken.

Biotechnology in Orbit

Just 30 years after Sputnik I went into orbit around an astounded world, in 1957, the first real orbital factory could be in operation, producing a stream of pharmaceutical products based on cells, enzymes, hormones and proteins. With the launch of the European Space Agency's Ariane L-6 in mid-1983, the competition really began to hot up for an orbital market which some predict could be worth £400 billion by the end of the century. But NASA's Space Shuttle had already gained an advantage in the race to the orbital factory by carrying into space the McDonnell-Douglas experimental electrophoresis test-bed, as part of the EOS (Electrophoresis Operations in Space) project. With factories standing idle around the world, this growing interest in orbital real-estate may seem illogical, but EOS illustrates some of the very real economic benefits which could flow from the eventual industrialisation of space.

The basic EOS device met all its objectives when it was flown for the first time on the fourth Shuttle flight, in June, 1982. To understand why these objectives were important, it helps to know something about the process of electrophoresis itself. It involves the movement of charged particles in a solution under the influence of an electrical field. Because particles of different charges and sizes move at different rates towards an electrode with an opposite charge, electrophoresis is especially useful for separating the different components of a mixture—such as a fermentation broth.

Of course, there is nothing to stop electrophoresis taking place here on Earth. For example, in 1983 I was shown the new Biostream machine at Harwell, which was about to be marketed by a John Brown subsidiary. Whereas previous efforts to develop production-scale electrophoresis had foundered on the problem of 'convective turbulence', which re-mixed the carefully separated fractions of a fermentation broth, the Harwell design resulted in a machine which could process several litres of liquid an hour—and with up to 29 separate fractions being drawn off simultaneously from the top of the system. So, if you mixed 29 differently coloured materials together, the transparent tubes through which the fractions are drawn off would look like a rainbow (except that the colours would almost certainly be in the wrong order). Electrophoresis, as performed in the Biostream process, offers unusually mild mechanical, chemical and thermal conditions, enabling sensitive, biologically active substances to be processed with minimal loss of activity. One application is in the preparation of the blood-clotting Factor VIII, with the electrophoretic separator giving higher yields and better purities than conventional precipitation procedures.

The type of electrophoresis commonly used in laboratories is called 'static' electrophoresis, because the mixture to be separated, the 'sample', is placed on a stationary, or static, medium like a porous gel plate. An electrical field is applied across the medium for long enough to allow the differently charged particles to migrate to separate areas of the gel. However, although good separation can be achieved with this technique, only a small amount of sample can be separated at any one time, typically about 0.01 millilitres—which explains why this approach is used for analysis rather than production. If, on the other hand, you want to go into commercial production, continuous-flow electrophoresis is what you need —and what both Harwell and McDonnell-Douglas have been developing. In this process, a stream of sample is continuously injected into a flowing buffer solution, called the 'carrier fluid', which carries the sample from the bottom to the top of the electrophoretic chamber. As the sample flows through the chamber, an electric field is applied across the direction of flow. As in static electrophoresis, this causes the differently charged particles to spread out into different particle streams that emerge through separate outlets at the top of the chamber.

The fly in this ointment is gravity, which seriously limits the extent to which sample solutions can be concentrated, thus limiting the output of even industrial-scale electrophoresis. Gravity also causes the convection currents which limit output and degrade purity. And, finally, it causes what is known as 'bandspreading', which likewise affects the purity of the fractionated output.

In the weightlessness of space, things are very much easier. The problems associated with limited sample concentrations, convection currents and bandspreading are virtually eliminated. Because there is no gravity to limit concentrations, most sample concentrations can be increased at least a hundredfold—and some by as much as 400 times. Furthermore, because gravity-induced convection currents are eliminated, a deeper chamber can be used, with larger inlets, permitting the injection of up to four times more sample.

The separation system which has now gone into orbit on a number of occasions looks relatively simple from the outside. There is a 'fluid systems module', where the separations take place; a microcomputer, used to monitor and control the separation process; a refrigerated module, where the materials are stored before and after processing; and a water-cooling module. But, having learned from earlier missions, McDonnell-Douglas made some significant changes in the way that the process is operated.

First, the voltage applied across the chamber was nearly trebled, from 140V to 400V, while the amount of time the sample spent in the chamber was also increased by 60 per cent. So, whereas the first flight resulted in the separation of 463 times as much material as would have been possible on Earth, the second time around this was boosted to over 700 times as much. The samples separated during the second mission were a standard laboratory mixture of rat and egg albumins; a cell-culture fluid containing many types of protein; and two samples of haemoglobin.

Clearly these results, coupled with the realisation that EOS could produce fractions which are up to five times purer than those from earthbound units, encouraged the company enormously. In 1984 one of its mission specialists, Dr Charles Walker, became the first paying industrial astronaut, borne aloft in *Discovery*. The company paid $80,000 as his fare. Meanwhile, plans were also afoot to put an automatic version of EOS, some 24 times larger than the earlier system, into orbit. McDonnell-Douglas and Ortho Pharmaceutical, a subsidiary of Johnson & Johnson, have been looking for high-value bio-products which could be separated in this way.

It is still far from clear whether such work will produce new cures for major diseases, or cheaper cures for diseases which can already be treated. Despite operational difficulties experienced on the maiden flight of *Discovery*, it had been thought that the hormone produced by the electrophoresis experiments run by Dr Walker would be active. It proved not to be. It looked as though bacterial contamination of the EOS system, which was aggravated by pre-flight delays, resulted in the destruction of the target hormones.

But the early successes of the EOS missions do suggest that sufferers from a number of important diseases might soon justifiably turn their eyes to the heavens in search of a cure.

ELEVEN
The New Services

Dig a hole in the ground and the chances are that someone will soon happen along and offer you some form of service: another pair of hands, a sharper shovel, or advice on where you should *really* be digging. Any new industry not only provides services, as biotechnology is beginning to do, it also consumes them. In this chapter, then, let's look at some of the diverse services for which the biotechnology industry has developed an appetite.

One of the first people likely to turn up beside your hole will be someone wanting to sell you insurance, offering to cover you against the risk of cutting through your ankles or someone else's gas main. And, true to form, insurers have been developing new forms of cover for the emerging biotechnology industry. The Sun Alliance Insurance Group, for example, teamed up with Reed Steenhouse to cover a wide range of risks, including any legal liabilities resulting from professional negligence.

So, in the knowledge that you are fully protected, you need lose no more sleep for fear that accident or theft might rob you of your hard-won cell lines or cultures. Some cells may also undergo spontaneous genetic change which strips them of all your carefully engineered attributes: again, there is almost certainly an insurance policy covering this eventuality. You can insure against cross-infection between your cultures or the loss of your experimental animals. And you can cover yourself against the biological equivalent of a nuclear 'melt-down', with the insurer paying any legal costs resulting from contamination —and the costs of decontamination of your facilities and, if necessary, your neighbours'.

On the other hand, you may feel that insurance money would be no real compensation for the lost cell lines or cultures. In that case, you could take a sample of your favourites to your nearest culture-collection 'centre', and pay for them to be held for you. Culture collections are to microbiology what reference libraries are to other professions. Most countries have a number of culture collections, of varying richness.

Probably the richest is the American Type Culture Collection, a non-profit organisation based in Maryland. It houses the most diverse collection of strains of algae, bacteria, bacteriophages, fungi, plant and animal viruses, anti-sera, protozoa and cell lines, plus a rapidly growing collection of plasmids, recombinant-DNA vectors and hybridomas.

The UK has national collections that cover dairying, industrial, marine, medical and veterinary bacteria; fungi of various types, including pathogenic fungi and those responsible for rotting wood; algae and protozoa; and yeasts. Some, like the National Collection of Industrial and Marine Bacteria, have been privatised. More modern facilities are also now appearing, like the National Collection of Animal Cell Cultures, which opened its doors in July, 1984, and is based at the PHLS Centre for Applied Microbiology and Research (CAMR), Porton Down, Salisbury.

Such animal cells are used to produce monoclonal antibodies, enzymes, interferon and other therapeutic substances. Dr Alan Doyle, curator of the new CAMR cell-culture collection, plans to investigate how cells change in storage. Some cell lines may be stored for as long as 30 years and there is a real possibility that, however carefully they are managed, they may lose their carefully engineered characteristics over time. Any such research will aim to make the collection's services more useful to users. Unlike many of the older collections, which are not really set up to respond to industry's needs in the 1980s, these new units are very much more commercial. Some are also beginning to computerise. The UK's National Collection of Yeast Cultures, for example, has fed information on 100 separate characteristics of each of the 2,000 authenticated yeast cultures it holds into a computer. Anyone can pay about £20 to run a search for yeasts offering a specific combination of characteristics. In the USA, GenBank, a data-base sponsored by the US National Institutes of Health, the National Science Foundation and the Departments of Energy and Defense, is now open to subscribers worldwide. It holds over 1,500 nucleic-acid sequences of 50 or more nucleotides. There are many other data-banks in operation, with more coming on-stream all the time.

The Bug Hunters

And, just as fleas have smaller fleas upon their backs to bite 'em, so service industries themselves consume services. Culture collections need people to collect new strains for them, for example. A key ingredient of Japan's success in many areas of biotechnology has been the enormous effort it has put into screening wild microbes for novel enzymes and other potential money-spinners. Kyowa Hakko, for example, developed one of its leading antibiotics, Sagamicin, from micro-organisms found in soil taken from Sagami Lake.

At first it looked as though genetic engineering might replace much of this screening, but that now looks a very remote possibility. Indeed, even in Europe, many biotechnologists bring back soil samples from their business travels or holidays. We have seen how companies like RHM found genetic treasure almost literally under their own doorstep. Another UK company, Tate & Lyle, discovered one of its most exciting potential products in a rhubarb patch.

The contribution which such 'bug hunters' can make is strikingly illustrated by the story of the discovery of a powerful new tool in the hostile conditions of a natural salt lake. The new tool, a restriction enzyme called *AhaIII*, was found by Dr Nigel Brown and Philip Whitehead of Bristol University. They found it in a blue-green alga, *Aphanotece halophytica*, which was being investigated by Professor Walsby in Bristol's botany department. The alga is found in only a very few places, such as the Dead Sea and the salt lakes of the Sinai Desert. It thrives in water containing upwards of 10 per cent salt at temperatures reaching 60–65°C (140–150°F).

P & S Biochemicals, the Liverpool company which brought *AhaIII* into mass-production, is highly skilled in what one of its directors, Dr Peter Dean, has dubbed 'biochemical fishing'. Dean, also involved in the Agricultural Genetics Company (see page 116), used affinity chromatography to hook restriction enzymes like *AhaIII* out of the 'protein pool' extracted from algae and other sources. This process used to be very slow, but has been accelerated by a new technique, fast-protein chromatography, which has cut the time needed to screen a culture for a desired protein from several days to a mere 15 minutes.

The most exciting thing about *AhaIII*, however, is not the way it was hooked but what it does. It enables genetic engineers to cut the 'zip fastener' structure of the DNA double helix in a totally new way. This single product, Dean recalled, 'took us into the 20th century, and we went in with a bang. The result was more or less instant fame and the company grew by a factor of four every year.'

AhaIII turned out also to be astonishingly stable. 'When we sent the first commercial batch to Genentech,' Dean noted, 'it was exposed to summer temperatures on the floor of the New York post office for a week, which must be horrendous—it's literally cooking it. When it arrived at Genentech, there was water leaking out of the box. They thought it might be a bomb and let it incubate. That's the sort of stability testing *AhaIII* had, quite unintentionally.' It passed with flying colours.

P & S are now encouraging bug hunters around the world to send them samples. Indeed, a single article in the *Financial Times* hooked about ten microbial contributors—'and they've got about 6,000 bugs a year for us to screen', said Dean. 'We're getting bugs from all over the place, we're getting them from California through to Moscow. We've got a lot out of Windermere, a lot out of North Wales, a lot out of the Aberystwyth area.' Biotechnologists, in short, may soon be claiming their holidays against income tax.

Ivory Towers in the Marketplace

Dean, in fact, is a prime example of the new breed of entrepreneurial academic, common enough in the USA but still something of a rare species in Europe. His early links with P & S caused ructions with his colleagues in the biochemistry department at the University of Liverpool, who felt he should not be using university resources for commercial purposes, although he enjoyed strong support from the top level of university management. Ironically, in the early days P & S had made more money from award schemes promoting new enterprise than from its business activities, but the money enabled it to set up the production facilities it needed, bypassing the somewhat precarious relationship with the university.

The demand for the services of academic bioscientists and biotechnologists often seemed limitless, however, despite the intense controversy which surrounded the relationships between industry and the groves of academe. One of the most controversial link-ups involved the West German company Hoechst. Its decision to go for a ten-year $70-million biotechnology agreement with the Massachusetts General Hospital, rather than investing in West German research, caused an explosion of protest in its home country. However, several of Hoechst's competitors had soon signed agreements with other West German centres of excellence: BASF promised the University of Heidelberg $450,000 a year over five years, while Bayer Leverkusen signed a collaborative agreement with the Max Planck Institute for Plant Research.

When the US Office of Technology Assessment (OTA) published its international analysis of commercial biotechnology in 1984, it concluded that 'the United States has the most effective and dynamic university-industry technology transfer process of the six countries examined'.

It also predicted that

> US university-industry relationships in biotechnology will most likely follow the same pattern that they have in other high-technology areas. First, scientific breakthroughs generate a period of hyperactivity in university-industry relationships. This hyperactivity phase is characterized by the promise of 'big bucks', which leads to a short-term faculty and postgraduate drain. After the industry goes through its initial phases, an equilibrium state is reached and a fairly healthy symbiotic relationship emerges.

Which is not to say that there is no friction: there is. The University of California, for example, requested three of its professors to terminate or alter research contracts with industry in 1983, on the grounds that there was an existing or potential conflict of interest. One of these contracts was in the genetic-engineering field: Milton Schroth, professor of plant pathology on the university's Berkeley campus, was principal investigator in a research contract with Advanced Genetic Sciences, based in Connecticut. The $82,000 contract was taken over by one of Schroth's colleagues.

Japan, which the OTA picked as the most important emerging competitor to the USA in biotechnology, has confused many who have tried to understand its university-industry relationships. The violent student revolt of the 1960s gave rise to an acute sensitivity about such relationships, but they turn out to be very much stronger than an outsider might expect having heard the protestations of Japanese professors about their independence from industry in any guise.

Although there are strong rules against formal links between industry and the country's elite 'national' universities, such as Kyoto, Osaka and Tokyo, individual professors have much greater opportunity to work with industry. There is also a key distinction between basic and applied science departments at Japanese universities. The professors who tell you that they are totally untainted by industrial ties tend to be working in biology, biochemistry and medical science; those working in applied science, such as bioprocess engineering, have much stronger industrial contacts. They also spend a great deal of time ensuring that their students find jobs in industry . . . and keep in touch thereafter.

But, the OTA pointed out, 'Japanese professors at the national universities are forbidden to take other positions simultaneously with their university work, and all donations towards their research must be made through formal university channels. No fees for consulting can be accepted, and offers of stock options are unheard of.' Clearly, a radically different situation from that prevailing in the USA. Japanese professors, moreover, were not permitted to hold patents or receive royalties until 1981.

The UK, renowned for the weakness of technology transfers from universities or public-sector research institutes to the country's own industry, has been working hard to redress matters, with a degree of success. Celltech was set up to exploit research funded by the Medical Research Council, while the Agricultural Genetics Company aims to do the same for work funded by the Agricultural and Food Research Council. In some universities, such as Sheffield (see page 122), new companies have been set up to commercialise a particular department's research. Elsewhere, new centres have been set up to act as a more general bridge between a university—such as Cambridge (page 52) or University College, London—and industry.

The Leicester Biocentre, based on the University of Leicester, was a case of second time lucky. An earlier plan foundered because of the cost of providing new laboratories and disagreements over royalties. Ultimately, it was decided to re-equip an existing suite of laboratories at the University, with backing from five major companies: Dalgety Spillers, Distillers, Gallaher, John Brown and Whitbread. The Biocentre is working on the gene structure of yeasts and higher plants.

Although the Biocentre's first director, Professor Barry Holland, predicted that it would be 'unique in Europe as a research centre specifically designed to foster university–industry collaboration', this was rapidly becoming a prime objective throughout Europe. In France, for example, companies like Elf Aquitaine and Transgene were locating their new biotechnology facilities in close proximity to university centres of excellence—Toulouse and Strasbourg respectively.

On a very much smaller scale, some universities which have invested in expensive biotechnology equipment have been defraying part of the cost by offering services to outside users. For example, a new Scottish company, Sequal, offers a contract sequencing service for peptides and proteins, using a system bought from Applied Biosystems. Sequal draws heavily on the expertise and resources of its directors, who work at the University of Aberdeen.

Indeed, a striking side-effect of the swingeing cuts in university funding in the UK has been the movement of many academic centres of biotechnological excellence into the marketplace. Imperial Biotechnology, launched in 1982, illustrates the trend. Following funding cuts, Imperial College, London, was forced to consider shutting down the battery of fermenters it had built in the 1960s with the aid of a £500,000 grant from the Wolfson Foundation. Several of the fermenters had already been modified to provide complete containment, permitting the cultivation and handling of genetically engineered micro-organisms on a pilot-plant scale, but the annual running costs of about £300,000 were proving something of a strain. What to do?

The unit had carried out contract work for a number of major industrial clients over the years, including Beecham, ICI, RHM, Tate & Lyle, and Wellcome. Imperial had worked with ICI on polyhydroxybutyrate (see page 180), the biodegradable plastic made by bacteria, and with RHM on mycoprotein (see page 152). It had also carried out the first fermentations of Biogen's genetically engineered *Escherichia coli*, providing Biogen with useful quantities of interferon in advance of its own fermentation operation being set up. But the unit had been operating at only about 30 per cent of its fermentation capacity—and perhaps 10 per cent of its separation capacity. So the decision to market aggressively that spare capacity made a good deal of sense.

These are just the sort of services that the biotechnology industry needs—and will pay for. Very few companies are in the position of Du Pont, which announced that it would spend nearly 20 per cent of its $1-billion 1984 R&D budget on what it considers basic research, up from only 10 per cent five years previously. Lesser companies will be much more dependent on what they can extract from other people's research, often carried out in universities. Equally important, although critics claim that such tie-ups will force universities to concentrate only on work with some immediate prospect of profits, these new links are injecting much-needed adrenalin into university and other academic departments, which would otherwise have been in danger of falling behind the times. At the same time, the new companies now being set up will provide a demand for new generations of university graduates—further supporting the relevant departments in universities and other institutions. While there are inevitable problems, with growing constraints on the free communication of some forms of information, this trend must be a welcome one for all concerned—except, possibly, the public.

The Enzyme Tailors

The more resources academic bioscientists and biotechnologists can lay their hands on, the more likely it is that they will have key services to offer industry. Many of biotechnology's healthiest companies, in fact, are those which have developed strong service businesses. Novo Industri, of Denmark, founded in 1925, is a leading example. Novo's business was built on the back of its skills in extracting and purifying two types of compound from animal tissues: hormones, including insulin, and enzymes, such as trypsin. We have seen how Novo and Genentech emerged as major competitors in the insulin market (page 81), but Novo dominates world markets in another area: enzymes.

By the early 1980s, Novo's enzymes had achieved a 50 per cent share of the world market, an impressive achievement by any standards. Enzymes, as we have seen, are proteins found in all living materials: they control biological processes and chemical reactions, in both the synthesis and decomposition of complex molecules. While enzymes can be extracted from animal and plant materials in small quantities, considerable progress has been achieved in the isolation and subsequent fermentation of enzyme-producing micro-organisms.

The detergent industry was one of the first major industrial users of enzymes, with sales to this industry still accounting for nearly half of Novo's total enzyme sales. These enzymes dissolve protein-based stains and offer energy savings, boosting the cleaning power of cold-water detergents. Unlike most other industrial enzymes, however, detergent enzymes are part of the final product, rather than simply being used in the production process—and therein lies a potential problem: they come into contact with consumers. Allegations were made in the early 1970s about health hazards, triggering a debate which cut Novo's sales in this area from $48.7 million in 1969 to just $8.5 million in 1971. The US Food and Drug Administration concluded in 1971 that there was no substantial evidence of a consumer safety problem and, after some product reformulation to cut down on dustiness, Novo's detergent-enzyme sales have climbed back to 1969 levels—although the recovery in US sales has been slower than elsewhere.

These concerns re-emerged, however, when Unilever launched its New System Persil Automatic washing-powder, with a £7-million advertising campaign, in 1983. Within just a few months over 5,000 letters were received from eczema sufferers whose condition had allegedly been worsened after wearing clothes washed in this new 'biological' powder. The withdrawal of an older, enzyme-free powder, Persil Automatic, had left the UK's five million eczema sufferers without a nationally available powder free from both enzymes and other low-temperature additives—the ingredients thought to have caused most of the irritation reported, ranging from itching to inflamed sores. Unilever later reintroduced its enzyme-free powder.

The starch-processing industry is another major enzyme user, accounting for over a quarter of Novo's total enzyme sales in recent years. The primary use of enzymes here is in breaking down starch molecules into various sugars, including glucose and fructose, for use as sweeteners—an application in which they were increasingly favoured as raw-sugar prices became ever higher and more volatile. Starch processors also use enzymes in the manufacture of ethanol, although even the US gasohol programme, considerably cut back by the Reagan administration, has not yet resulted in significant sales as far as Novo is concerned. During 1981, however, it introduced a new enzyme, Spirizyme, for use in ethanol production.

The wine and fruit industries use Novo enzymes to improve juice yields, colour and clarity. Although its enzymes have so far been used primarily in the processing of grapes and berries, Novo has developed enzymes for use with apples and other fruit with a high starch content, and for citrus fruit. Cheese production has been another major business area: in response to a rennet shortage in 1965, Novo introduced its enzyme Rennilase, which can be mass-produced for use in milk coagulation and cheese production. Other, more fragmented, enzyme markets include brewing, textile production, pharmaceutical manufacturing, leather tanning and baking.

And, as an example of services built on services, consider the business strategy of one of the UK's latest biotechnology companies, Biocatalysts. Anyone who read the news that Grand Metropolitan, the UK food, beverages and leisure group, was pulling out of Biogen and assumed that Grand Met was pulling out of biotechnology altogether jumped to the wrong conclusion. Biogen's accelerating move into pharmaceutical markets (see page 70) simply made it less attractive for a food and beverages investor. Grand Met's continuing commitment to biotechnology was underscored, within days of the announcement that it was withdrawing from Biogen, by the launch of Biocatalysts, its first biotechnology subsidiary.

Recruited from Novo's UK operation, where he was technical manager, Tony Godfrey became Biocatalysts' managing director. In early 1984 he told me:

Enzymes already have a relatively small number of well tried and effective applications on the industrial scale, and we accept that there are many more that can be brought on-stream if there is a much closer dialogue between enzyme technologists and industrial users of enzymes. Biocatalysts will supply *tailored* enzyme preparations for specific and optimum use, supported by services aiding applications development through laboratory and pilot-scale trials. Our services also include an extensive international index of data on the range of commercially available enzymes which may be interrogated by industrial and academic clients.

In fact, the enzymes market looks like being one of the key biotechnology growth areas, in European as well as other major industrial nations, such as Japan and the USA. The West European market alone is projected, in a survey produced by market consultants Frost & Sullivan, to grow from a $46-million value in 1982 to $89 million, in constant dollars, by 1992. Over the same period, the US market is expected to grow from $30 million to $88 million.

Significantly, however, a very small number of enzymes (Frost & Sullivan says 16) out of a total of around 200 which have been thoroughly investigated for industrial use account for an astonishing 90 per cent of the West European market. The main problem has been one of cost, with the major enzyme producers (Novo included) unwilling to develop products for markets which do not offer multi-million dollar annual sales.

Godfrey's experience, while working with such companies as ICI, Biocon and Novo, had convinced him not only that the enzymes business is likely to remain a key area in biotechnology, but also that some of the fastest growth may well happen in markets whose relatively low value, on current criteria, means that they are largely ignored by the larger enzyme producers. But, and here is the significant point, Biocatalysts does not see itself setting out to challenge the major existing producers in their core, high-volume markets.

The company's focus, Godfrey stressed, 'is not as obvious as small-volume-rather-than-large-volume. It is much more a case of off-the-peg versus tailored enzymes. In many cases, even users of off-the-peg enzymes are now talking about wanting to "tidy up" those enzymes, to close in on the precise targets which they have. In other cases, they simply want to add on a piece of processing.'

The computerised index of enzymes which Biocatalysts offers

is also a canny ploy for identifying potential customers for the company's mainstream services. 'If a customer comes up with a question we can't answer from our ENZIDEX data-bank,' Godfrey explained, 'then we'll know that, one, he wants it and, two, it isn't available. So we have a potential customer.'

Profits Down the Drain?

Enzymes have also been a central interest at Genex which, unlike such rivals as Biogen, Cetus and Genentech, has been steering clear of most pharmaceutical targets. It accepts that most recent advances in genetic engineering have been associated with pharmaceutical applications, but its own strategy emphasises areas of biotechnology where commercialisation can rapidly follow the completion of the R&D process—rarely the case with pharmaceuticals, which require lengthy, expensive testing and can hit unsuspected regulatory snags.

One unusual product which Genex brought to market in 1984 was a hydrolytic enzyme, trade name 'Proto', which had been designed to unclog drains. Genex had used genetic-engineering techniques to construct strains of *Bacillus subtilis* that expressed the enzyme alkaline protease at a high level. The market for effective drain cleaners is substantial, but existing products such as lye (sodium hydroxide) and bleach (hypochlorite) are noxious and hazardous. The Genex enzyme is much safer and, unlike rivals which clump and shrink hair that is clogging drains, allowing it to be pushed clear, Proto dissolves the hair. The question is: can it work fast enough? Shelf-life was another worry, with the enzyme breaking down too quickly in storage, but Genex thinks it has cracked the problem with a method for producing beads of the enzyme and the other active ingredients.

Genex was also one of the first companies to develop training services for would-be biotechnologists, particularly from Japan. 'The Japanese are interested in moving fast,' explained Genex president Dr Leslie Glick, 'and we might as well make money doing it for them.'

An early jibe aimed at the smaller genetic-engineering companies was that they were in danger of becoming little more than 'gene boutiques', pursuing customers with services tailored to the shifting whims of fashion. But Genex used its early service-oriented days to good effect. Glick estimated that the company had looked at over 600 products for clients by early 1982. In doing so, it had built up an excellent idea of where attractive markets were likely to develop. One area in which it is convinced that biotechnology will be of considerable service is environmental protection. This is more than a little ironic, given the way that some environmentalists have reacted to biotechnology. It is worth recalling briefly just one of the cases where the plans of biotechnologists sent a number of environmentalists into near-apoplexy.

Snow to Go

Artificial snow, another out-of-the-ordinary bio-product, had been one spin-off from research on the use of genetic engineering to help cut frost damage to crop plants. 'It's not your typical high-technology product,' agreed Daniel Adams, chairman of Advanced Genetic Sciences (AGS), the same company Professor Schroth had advised. The heart of the company's business is a bacterium, *Pseudomonas syringiae*, which has been shown to accelerate the formation of ice crystals and snow. But its first attempt to carry out field trials of a *Pseudomonas* product caused an environmental furore in the USA, largely because the proposed experimental trials would have involved the first intentional release of genetically engineered organisms into the open environment.

Scheduled to start late in 1983, in a potato patch near Tule Lake, California, the trials were meant to evaluate the ability of genetically reprogrammed bacteria to control frost damage. To understand why the sponsors of the trials, who included the University of California at Berkeley, felt that it was worth running the environmental gauntlet, it helps to know a little more about the bacterium itself.

Tap water, as some of us know to our cost, freezes at 0°C (32°F), while distilled water can drop to - 15°C (5°F) unless it contains impurities to act as 'seeds' on which ice crystals can begin to form. These small particles trigger the process of ice nucleation, with some particle shapes seeming to be dramatically better at promoting it.

Unfortunately for farmers, one of the proteins found on the surface of *P. syringiae*, the most commonly found bacterium on plants, is so adept at ice-nucleation that it could almost have been designed for the task. Dr Steven Lindlow, a plant pathologist at the University of California at Berkeley, was one of those who began to wonder whether the bacterium could be responsible for a significant share of frost damage to crops. If so, and if farmers could be persuaded to part with money for the service, there might be a useful market for genetically tailored bacteria which could shoulder out their troublesome cousins.

The damage occurs not just as a result of chilling but also because ice crystals form inside the plant tissues, disrupting their structure. Normally, this damage becomes significant at around −2°C (28.4°F), but it has been found that, if the plants are cleaned of the bacteria, they can cope with frosts as severe as −10°C (14°F). The Berkeley research team had pinpointed the gene coding for the ice-nucleation protein, and Lindlow and Dr Nicholas Panopulos had removed it from some of the bacteria, as a first step towards producing large cultures of the mutant strain, known as 'ice-minus'. Next, the team planned to spray the ice-minus bacteria on potato plants to see if they dislodged the natural, ice-nucleating bacteria and thereby cut frost damage.

Although other scientists believed that a more effective approach would be to use viruses to destroy the bacteria, without resorting to genetic engineering, the National Institutes of Health Recombinant-DNA Advisory Committee cleared the trials. But then Jeremy Rifkin, one of the more vociferous and effective opponents of genetic engineering, sought a temporary restraining order from the US District Court, claiming that the bacteria could spread, decimating California's wildlife and disrupting weather patterns by affecting the natural processes of condensation in the atmosphere. And Rifkin had another card up his sleeve. He suggested that an environmental-impact study should have been carried out under the terms of the National Environmental Policy Act. Although not a scientist himself, Rifkin had an ally in Dr Peter Raven of the Missouri Botanical Garden in St Louis. Raven filed an affidavit stating that, if the bacteria did spread away from the trial area, the modified genes could also spread by natural transfer processes to other plants, making them more frost-resistant, lengthening their growing season and thus affecting all the species dependent on—or otherwise affected by—them.

In the event, with the frosts due to start at the end of October, the university postponed the trials until the following year. Stalled on this front, AGS pressed ahead with another application of *Pseudomonas*. In this case, the company had been working with the University of Wisconsin and the resulting snow-making technology looked promising. Conventional snow machines are used by many ski resorts to ensure that skiers have something like snow even when natural snow is in short supply. The machines, which typically use about 180 litres (40 gallons) of water a minute, use a compressed-air gun to shoot water droplets into the air. If it is cold enough, they come down as ice crystals. Unhappily, they work best at temperatures around −7°C (19.4°F) and a great deal less so as the temperature begins to rise towards 0°C (32°F). This is where the bacteria come in. At −7°C (19.4°F), only about 16 grams of the bacteria are needed to turn every 10,000 litres (¼oz per 1,000 gallons) of water into artificial snow. Even when the temperature is 0°C (32°F), *ersatz* snow can be made, provided 560g of bacteria are added to every 10,000 litres of water (9oz per 1,000 gallons). AGS pointed out that the bacteria enable snow machines to deliver 'a much higher ratio of snow to water' at any temperature, although one of its research scientists warned that the end result might not be 'nice, lacy snowflakes'. But, he pointed out, lacy snowflakes 'aren't all that common out in nature'.

Even so, the operator has considerable control over the quality of the artificial snow. The more bacteria used, the drier and fluffier the snow is going to be. And, for environmentalists, it is worth stressing not only that the bacteria are *dead* before they are injected into the snow machine, but that they are a wild strain, which can be found widely dispersed in the environment.

AGS hopes that resorts will use bacterial snow to extend the skiing season and to fill in during warmer spells. But, just in case this idea does not take off or receives an unexpectedly hostile reception from the environmental lobby, the company has been thinking of alternative applications. One involves using the bacteria instead of silver iodide as a cloud-seeding agent, to bring rain to drought-stricken areas. Another is to go after the home-insulation and cooling markets. Many US homes have snow-filled attics, both to provide insulation and to drive air-cooling systems through the summer months. If bacteria are used to make this snow, AGS believes, it should be possible to alter its texture in such a way that it lasts longer. This looks like an attractive market, not least since Rifkin and his allies will presumably be less likely to complain about what people do in the privacy of their own attic.

Thriving on a Diet of Poison

Many environmentalists might be surprised to hear about the environmental services which biotechnologists are already providing. In retrospect, the fact that the first patented genetically engineered microbe was tailor-made to relish and neutralise an important pollutant (see page 32) can be seen as something of a portent.

Indeed, a number of recent market surveys have suggested that the environmental applications of biotechnology will represent one of the fastest growing bio-sectors. One such study, by Business Communications Co., predicted that, while the total

US market for microbes and enzymes will grow at about 7 per cent a year to 1991, environmental clean-up applications will grow at an average rate of 17 per cent. These start from a smaller base, of course, but are seen as particularly important in the wake of such disasters as Love Canal and the Valley of the Drums. Equally important, Genex would tell you, less capital is needed than for, say, food applications, and government regulations tend to be much less of a problem.

One area in which biotechnology is beginning to have a significant impact is toxicity testing. With an estimated 70,000 synthetic chemicals being traded, of which perhaps 25,000 are in common use, there has been growing concern about the possibility that many of them may be causing cancers, genetic defects and reproductive failures. Biotechnology is now turning out some interesting new tests for detecting some of these effects.

When the US National Academy of Sciences and the Environmental Protection Agency published a report they had sponsored on mutagenicity tests for chemicals, they reviewed about 50 such tests, including many using bacteria, fungi, mammalian cell cultures and insects. Some were cheap, quick and accurate, but a great deal more work needs to be done in this area.

Dr Ananda Chakrabarty, meanwhile, having been responsible for that first patented microbe, has not been blind to the opportunities offered by the backlog of persistent pollutants which are constantly coming to light. These chemicals need no further testing: we know they can be killers. Combining old breeding methods with new genetic-engineering techniques, he and his colleagues at the University of Illinois announced in 1981 that they had created a bacterium which lives on a diet of toxic chemicals, including the herbicide 2,4,5-T. Confronted by soil containing a relatively high concentration of this persistent herbicide, the bug will (under laboratory conditions) eat up to 98 per cent of the 2,4,5-T. Chakrabarty has been trying out his new super-bug on areas where the US Air Force carried out target practice with 'Agent Orange', which was used extensively during the Vietnam War defoliation programmes and contained 2,4,5-T. He took samples of bacteria from Love Canal, Elgin Air Force Base in Florida and an Arkansas dump, hoping that natural selection would already have produced some fairly tough microbial survivors. It had.

He then extracted plasmids from the toughest survivors, which carried genes coding for enzymes capable of breaking down toxins, and inserted them into *Pseudomonas* bacteria. Next, he placed his new creations in a laboratory tank and, over time, exposed them to increasing levels of 2,4,5-T. Eventually, the survivors were eating 2,4,5-T as their staple diet.

Some companies, like H. P. Bulmer and ICI, have developed biological means of pollution-control as spin-off products from their core research. Bulmer, which makes cider, found a bacterium which thrives in hot (60–70°C), acidic effluents. The company's £24,000 digester has been saving it £30,000 a year. ICI, on the other hand, took advantage of naturally occurring enzymes found in fungi, which use the enzymes to counter cyanide released by damaged plants on which they are feeding, to treat cyanide in a laboratory-scale experiment—and found it worked. ICI, however, is not in the pollution-control business, and its earlier attempts to market pollution-control systems it had developed, like the Deep Shaft and Flocor systems, ultimately proved abortive, with the relevant activities spun off to smaller companies. But other outfits, like Polybac, based in Pennsylvania, have made such technologies their core business.

The principal subsidiary of Cytox, Polybac keeps some 450 strains of microbe in long-term storage. Polybac was founded by Thomas Zitrides, who had been struck by the fact that, while a great deal of attention had been paid to the mechanical engineering aspects of effluent-treatment systems, little was being done to improve the 'primary workers' in the treatment plants, namely the microbes which break down the effluents.

The first product he developed was Polybac itself, an improved bacterial strain for injection into treatment plants. Soon his list of products included others such as Lignobac, Oleobac, Nitrobac, Petrobac and Agrigest, the product names indicating the wastes that the particular microbes had been selected to tackle. After working through the same selection routes adopted by Ananda Chakrabarty, Polybac sells the final product in the form of a powder.

As in the frost-control experiments, however, there has been concern that the engineered microbes could escape from effluent-treatment plants and ravage the surrounding countryside. So biotechnologists have been working on microbes which will either consume all the waste they are given and then obligingly starve to death, or which have been given a fatal gene to limit their life-span. But critics like Professor Gordon Orians of the University of Washington's Institute for Environmental Studies are concerned that there might still be a problem: these short-lived bugs might pass on their defect to wild cousins by natural processes, resulting in widespread ecological disruption.

The Worm's Turn

New companies, including Biomechanics (page 166) and BioTechnica (page 167), continue to move into the environmental field. One of the most unusual is British Earthworm Technology (BET), which recruits its work-force in the dung-heaps of the world and is creating a profitable business out of some highly unpromising (and highly polluting) ingredients. It aims to commercialise worm-farming and 'worm-working' techniques developed at the Rothamsted Experimental Station, with support from the Agricultural and Food Research Council. But is the use of worms to turn wastes into saleable compost really *biotechnology*?

BET research director Dr Clive Edwards says it is. 'Earthworms feed mainly on micro-organisms growing on such organic wastes,' he notes, 'and in the process of feeding they fragment the waste, thereby increasing its surface area and encouraging further microbial activity.' Indeed, it has been found that some fairly sophisticated biochemistry is involved.

'The combined action of earthworms and micro-organisms turns much of the nitrogen content of the waste into the ammonium or nitrate forms readily available to plants,' he continues. 'At the same time, it also increases the amounts of phosphorus, potassium and magnesium in available forms. As the waste breaks down, its particle-size decreases progressively, and its moisture-holding capacity increases—often to a level similar to that of peat.' These effects, and others, make the resulting compost a particularly valuable plant-growth material.

Apart from trials on a wide range of animal manures, BET's worms have been tried out with considerable success on brewery waste, paper-pulp waste, potato waste and spent mushroom compost. The quantities of material available for treatment are very considerable: Ranks Hovis McDougall, for example, produces some 40,000 tonnes a year of spent mushroom compost, while Bowater Paper produces 30,000 tonnes a year of pulp wastes which it pays to have landfilled.

Once the worms have wrought their magic, they are converted into high-protein animal feed. BET estimates that the market value of the worms is £350–£400 a tonne, although in certain specialised applications, as in the feeding of young eels, where they are a prime food source, their value may be £2,000–£4,000 a tonne. If ten per cent of the available waste material were to be treated in the UK, assuming a figure of £400 a tonne of worms, the market could be worth £160 million a year. If the *compost* were to be priced at £80 a tonne, which BET has suggested may be conservative, then this market could be worth £1.8 billion per year.

But perhaps one of the most exciting longer-term applications of biotechnology will be in the field of environmental conservation. There is now a strong possibility, for example, that biotechnologists will help us save major areas of the world's vanishing rainforests from the destruction which is overtaking them. If these rainforests, the world's most diverse and productive ecosystems, are to have a future, new ways must be found to squeeze an income from them without destroying them in the process. Bioresources, a new UK company of which I am a director, is already applying some of the very latest techniques afforded by biotechnology to screen rainforest plants and organisms for potentially useful products.

Biotechnology, to sum up, is already spawning an extraordinary array of new species, few of which would have been predictable even a few years ago. Totally new horizons are opening up and biotechnologists are promising to solve most of our most pressing problems. Utopia, it often seems, is not only achievable but imminent. Is this a fair picture of our future?

Regretfully, one has to conclude that it is not. Not to put too fine a point on it, some biotechnologists seem hell-bent on beating their new plough-shares into swords.

TWELVE

The New Horizons

Imagine that you have $250,000 sitting in the bank, doing nothing but earn interest. Imagine, too, that you have just read an article on the new horizons which biotechnology is opening up, an article which has convinced you that the money should be invested in a new company which will develop exciting new products based on biotechnology. Assume that you have already quizzed many people in the industry to find out what the rather more way-out opportunities might be and that you have prepared a short-list of possibilities. Let's run down the list.

Of course, since this is the sort of information for which investors will pay good money, we shall keep the details to ourselves, but one item leaps at you: *Aphrodisiacs.* You hire a consultant to look into the possibilities and the resulting report does not dismiss the idea out of hand. You decide to call your company Sextech, and you get a designer to work up a letterhead and some business cards. Someone suggests that you call your product range 'Instant Passion' and initial soundings in the venture-capital community indicate that other people would be prepared to invest, too.

If this sounds like science fiction, be reassured: the science which could underpin a business like this is already being done. It has been known for many years, for example, that mating insects track down their partners thanks to powerful chemical sex-attractants called 'pheromones'. The quantities of pheromone involved are typically very small: even when in full production, a female insect may release her pheromones at the economical rate of just one gram per 23,000 years (1oz/652,000 years).

Today new pesticides are being developed which are a mixture of pheromones and insecticide, the idea being that the insect will come to the insecticide, rather than the farmer having to chase it. Another approach, which ICI has developed to control the Pink Bollworm, a pest which ravages cotton crops, is to treat the whole crop with the pheromone gossyplure, so that, while males may wing in from far afield, they cannot track down receptive females in the pervasive chemical 'fog'. Since the pheromones achieve the desired effect in astonishingly small quantities, this approach has enormous attractions—as far as the farmer is concerned.

If your consultant has dug into the relevant literature, you may already have come across some even more impressive scientific work. We have already looked at the role of luteinizing-hormone releasing hormone (LHRH) in restoring the ability of some infertile women to have babies. By now, you may be sitting in front of a stack of research reports suggesting that this hormone could have some very much more exciting effects. Your business plan is beginning to gel.

The Ultimate Aphrodisiac?

Flicking through some of these reports you discover that LHRH was initially thought to be simply a chemical messenger which triggered the production of hormones which, in their turn, stimulated the sex glands. But research on rats whose sex glands had been removed has indicated that LHRH can work as a sex stimulant in its own right. 'We can take away a rat's ovaries and pituitary gland,' explained Professor Robert Moss of the University of Texas Health Service Center, 'inject as little as ten-billionths of a gram of LHRH into its brain, along with a minimum amount of oestrogen as a primer, and the animal will engage in sexual behaviour for as long as eight hours at a stretch.'

Many new companies have been launched on the basis of very much less convincing animal studies. But, you may ask, what about the hormone's effect on people? Rats may like the product, but they are unlikely to buy it. Well, as it happens, Moss carried out some research on human subjects, too. The results were not as dramatic as those emerging from the rat tests, but they indicated that genetically engineered aphrodisiacs could have a future.

You may now feel that the time is ripe to bring Professor Moss, or someone like him, onto the board of your new company. But, before you try calling him, it is perhaps worth recalling that LHRH is only one of a number of substances found in the human body which may prove to have the desired effect. It is also worth pointing out that our understanding of what goes on in the human brain has improved dramatically in recent years, with many other novel substances becoming available for further study. Some, like endorphin (see page 85), may provide a new generation of pain-killers; others, like the neuropeptide vasopressin, could form the basis of a memory-boosting drug. Are you sure that it's aphrodisiacs that you want to be in?

We have learned a great deal about the body and its biochemistry in the last ten years. As far as the brain is concerned, for example, our understanding has taken a quantum leap. Whereas scientists thought that just ten chemicals (the neurotransmitters) were responsible for the electrical messages flashed between brain cells, they now know that the situation is very much more complicated. They have been forced to rethink their theories on the ways in which the brain communicates with the body, largely because of the discovery that at least 36 other chemicals—the neuropeptides—play important roles in cerebral chemistry.

Many of these neuropeptides also appear to play different roles at different times and in different places. The opioid peptides, for example, help the body control the sensation of pain, but they appear also to have a role in many other bodily processes, from the control of blood pressure to the management of body temperature. This is despite the fact that, in the average brain, such substances are generally measured in dilutions of less than one part per billion—which made them nearly impossible to detect until the advent of such new techniques as gene cloning and DNA hybridisation.

Among the potential pharmaceuticals found in the brain are adrenocorticotrophic hormone (ACTH), cholecystokinin (CCK) and vasopressin. The Dutch company Organon has isolated a fragment of ACTH which can give old people a sense of contentment, while CCK seems to be the agent which makes us feel well fed at the end of a meal. It has also been shown that ACTH fragments can boost the short-term memory, enabling people to recall what happened just a few minutes ago much more clearly; while vasopressin seems to enhance the ability of experimental animals to learn and remember: if rats are injected with very small quantities of vasopressin they are very much better at remembering ways in which they can avoid electric shocks dealt out by the experimenter.

No one suggests that developing such pharmaceuticals is going to be easy. The fact that many of these neuropeptides have a multiple purpose is in itself slightly worrying. Vasopressin is a case in point. The evidence suggests that, were it to be taken as a memory booster, the person taking it could suffer from fluctuating blood pressure, on which vasopressin appears to have an important influence.

An even stranger possibility is that people may one day be

injected with a chemical which induces hibernation. Baron Maximillian de Clara, who has invested at least $5 million in a wide range of research projects, spent $2 million on research into the strange behaviour of the African lungfish. If drought threatens, the lungfish makes itself a snug cocoon under the mud and, breathing through an airhole, seems perfectly content to wait for up to five years for the next rains to come. 'At first I just wanted to know how the lungfish was able to sleep for five years,' de Clara recalled recently, 'but then I saw that some practical use might come from it.'

Scientists from West Germany's Max Planck Institute for Biochemistry had dug up lungfish in Chad and taken them back to the laboratory to study their brain chemistry for clues to the mystery. They found a substance that seemed likely to be the active principle—and, when they injected it into rats, it put the rats to sleep. The substance, dubbed 'amcurine', was obviously worth a patent application, and de Clara duly submitted one in the USA. It was held up for what the authorities described as 'security reasons', and de Clara was later told that it was felt that amcurine could be useful in interstellar space exploration.

The day of the NASA or CCCP Rip Van Winkle is a long way in the future, but scientists working on what they describe as HIT (for Hibernation Induction Trigger), extracted from hibernating squirrels and woodchucks, believe that there could be a shorter-term commercial application for such substances. Peter Oeltgen, of the University of Kentucky, demonstrated that HIT taken from squirrels could send rhesus monkeys to sleep. It also caused their body temperature to drop several degrees, while their heart-rate plummeted to half the normal 150 beats per minute. The question some scientists are now asking is: can HIT help slow the ageing process?

Around the world, research teams, including that run by Professor Brian Clark of Senetek (see page 102), have been trying to pin down the various factors which cause our cells to age. Interestingly, it has been shown that the red blood cells of hibernating animals do not take up iron in the way that normal, ageing red blood cells do, and that the ageing of the collagen which links the body's cells together (see page 103) likewise seems to slow down. Just possibly, there may be a cure here for some aspects of the ageing process, although it is more than possible that such substances will find their main application in some totally unsuspected field.

On the Trail of the Oncogene

Eventually, in 1983, de Clara got his patent for amcurine, but by
then his interests had moved on: he was funding work at West
Germany's Freiburg University on interleukin-2 and had set up
his own company, Interleukin-2. Given that most people are
more interested in the prospects of fighting off cancer than of
hibernating through the winter, it seems likely that de Clara
will do better with interleukin-2 than with amcurine in the
foreseeable future—if his new company can fend off the competition.

The race to find cures for cancer is rivalled only by the race to
find out why cancer happens in the first place. There is a
tremendous air of excitement in cancer research, not least because of the discovery of 'cancer genes', or 'oncogenes' (from
the Greek *onkos*, meaning 'mass' or 'tumour'). We now know
that oncogenes are present in all cells; over 20 have been found
so far, although their purpose is far from clear. When some
trigger switches them on, they start to churn out abnormal
amounts of protein which, in turn, can cause cells to become
cancerous.

Among the companies which have been tackling oncogenes
are Genentech, Hoffman-La Roche, Merck and Syntex, which
has set up a joint venture called Oncogen with Genetic Systems.
All of these companies know that there is a difficult road ahead,
not least because there are over 100 diseases which qualify as
cancers; but the new knowledge and gene-splicing techniques
now becoming available promise an accelerating series of breakthroughs. New information about the oncogene is constantly
coming through, while more and more cancer-causing agents
(carcinogens) which could switch on particular oncogenes have
been identified.

'The discovery that it is an oncogene that issues instructions
to the normal cell to behave abnormally is enormously liberating,'
said one MIT molecular biologist, Dr Robert Weinberg. 'What it
meant was that you could now look at the site in the cell where
the carcinogen insult [damage] might take place. No longer is
the cancer cell a black box where vague and mysterious things
beyond our view take place.' If the oncogene triggers can be
identified, there is obviously at least a possibility that vaccines
could be developed against a range of cancers.

These insights often stem from basic research conducted over
many years, and there is no shortage of targets for future research. James Watson, having discovered the double helix with
Francis Crick and spent 15 years as a professor of biology at

Harvard, is one of the leading scientists who have been carrying out basic research on cancer mechanisms.

Crick, on the other hand, followed up the double-helix victory with key work on the way that genetic information is coded within the double helix. He then turned to work on the ways in which organisms develop. To begin with, he focused on the question of how genetically identical cells in a fertilised egg begin to differentiate into the different parts of an embryo. Later, with a special professorship at the Salk Institute, he moved into neurobiology, looking at the way nerve cells form and operate.

One project which could provide vital information on such development processes has been conducted over the last 20 years at the Laboratory of Molecular Biology. The organism which Dr Sydney Brenner picked for the research is not particularly inspiring, a millimetre-long soil nematode known to scientists as *Caenorhabditis elegans*. This fast-growing worm is complex enough to have an elementary nervous system, but has only 1,000 cells in its entire body. After years of painstaking study, every stage of the worm's development from the egg upwards has been carefully mapped. The next step is to look for mutant worms, extract the genes responsible for any mutations, and study what exactly they are and how they operate.

So, in a way, it is rather as though Watson and Crick took us to the genetic equivalent of the moon, leaving the rest of the genetic universe still to be explored. The break-out from Earth was an extraordinary development, but only the first step on a long journey. 'If we look back in ten years' time,' Crick predicted while celebrating the 30th anniversary of the discovery of the double helix, 'what we'll find is that there are lots of products that we haven't even thought of.'

Stanford was not exaggerating when it called the Cohen-Boyer gene-cloning technology an 'uncommon invention' (page 36), but new genetic-engineering techniques have continued to pour forth from research organisations around the world. The Japanese, for example, are using lasers to break open cells for a fraction of a second and allow foreign DNA to slip inside. Other scientists are working on ways in which recombinant RNA might be used instead of recombinant DNA in genetic-engineering research, so that proteins could be made without the aid of living organisms. And others still are working on special genes called 'cellular enhancers', designed to persuade animal cells to produce prodigious quantities of particular proteins, mimicking the activity of antibody genes, oncogenes and viruses.

New proteins are constantly being cloned. Various scientists, for example, have reported cloning one of the four genes involved in producing natural tooth enamel, suggesting the possibility that future tooth fillings could be almost indistinguishable from the teeth they have been used to repair. The mercury-silver amalgam with which most of our teeth have been filled does not bond as well as it might, with the result that the dentist has to drill a large, bottle-shaped cavity to hold the filling in place. Natural enamel would be much stronger—and fluoride might even be built in, to make it more resistant to decay.

Indeed, the pace of new development is so great that it is almost impossible to predict what will be possible by the 50th anniversary of the discovery of the double helix, in AD2003. What sort of world will our children's children look out upon as they chew their way through their mycoprotein menus with their imperishable teeth?

Putting Bugs into Computers

It will be a world in which computers and bugs collaborate. Growing numbers of ever-more powerful computers are being used everywhere you go in the biotechnology industry. You find them in data banks, keeping track of the millions of DNA sequences now known and needing sophisticated analysis. The QUEST software released by IntelliGenetics, to take just one example, can tap into the world's growing number of computerised data-bases, searching through textual, sequence and restriction map information on plasmid strains or bacterial stocks, and automatically provide you with a print-out of the latest information on specific genes, scientific references or patents.

In the world's first 'computer-aided conference on biotechnology', computers were used to hook together more than 200 laboratories in Europe, North America, New Zealand, Japan and the Soviet Union, as well as in such developing countries as India and the Philippines. No voice-contact was possible between those 'attending' this conference—and the information-exchange took months. No one expects this approach to replace the face-to-face contact afforded by traditional conferences, but the exercise illustrated the way in which a worldwide net of biotechnology-information suppliers and users is evolving.

Computers also store information on the latest restriction enzymes and vectors. They run analytical searches of large data bases, and have been throwing up some unsuspected findings, including a hitherto-unknown relationship between human platelet-derived growth factor and an oncogene found in a virus which can cause cancer in monkeys, suggesting that the oncogene might cause cancer by triggering excess production of a growth factor. Incorporated in DNA synthesisers, computers and microprocessors help string together synthetic genes. Hooked into fermentation processes, whether in small flasks or giant fermenters, they monitor and control an enormous range of fermentation conditions, helping maintain sterile conditions.

The continuing cross-fertilisation between computer technology and biotechnology will inevitably lead in some unsuspected directions. Indeed, in the long-term future, biotechnology might actually be used to build some computers. The microchip has revolutionised the use of computer power, with a computer which 30 years ago would have occupied a large room now fitting onto a chip the size of a child's fingernail. But recent work on the 'biochip' suggests that all sorts of extraordinary things could happen if biology were ever to replace physics as the driving force behind the evolution of the computer.

The biochip lies at the heart of a new area of science, variously described as 'molecular electronics' or 'bio-electronics', which is emerging as biotechnology collides with micro-electronics. Computers, currently used to redesign biological molecules, could one day be built from such molecules.

But enthusiasts often mean different things by 'biochip'. At its simplest, a biochip may be a chip built from conventional semiconductor materials, but designed in such a way that it can operate inside the human body, or in some other biological environment. Biosensors incorporating such biochips might be used to monitor water pollution, for example, or to warn soldiers of impending gas or biological attacks on the battlefield. Another biosensor might be a glucose sensor which could be installed in the bloodstream of a diabetic patient to monitor blood glucose levels and trigger the operation of an insulin pump implanted inside the patient's body. This sort of biosensor might consist of a short length of platinum wire, one end of which would be attached to a membrane containing fragments of an enzyme. Ultra-thin wires would lead from the other end of the platinum wire to the insulin pump. A biochip inside the membrane would pick up the shift in electronic signals produced as the enzyme fragments reacted with glucose in the blood. The higher the glucose level, the greater the electrical potential across the membrane and the stronger the signal sent from the biochip to the insulin pump.

Alternatively, a biochip might measure the concentration of calcium ions passing through a membrane. Dr John Barker of Warwick University has suggested that, implanted in the human body, such a biosensor could monitor a patient's heartbeat. If anything went wrong, the biochip could beam an alert to the emergency services and switch on a pacemaker to support the heart until help arrived.

A significant problem facing biosensor designers, however, is the human bloodstream itself, which represents a hot and chemically harsh environment for sensitive micro-electronic circuitry. Poorly designed biosensors may trigger the body's immune response or, alternatively, simply 'fur up'. Today's pacemakers, for example, need surprisingly large batteries to overcome the gradual decrease in the heart's sensitivity to pulsed current. A truly bio-compatible interface between the pacemaker and the heart could greatly improve the life expectancy of both pacemaker and patient.

Scientists are also exploring ways in which synthetic interfaces might be grown into the brains of blind patients, as a prelude to hooking them up to some system which, like a video camera, would restore their ability to 'see'. It is over ten years since a UK scientist first showed that a blind woman could see bright flashes of light when her exposed brain was touched with an electrode. Since then, blind patients with electrodes implanted in their brains have recognised simple shapes, letters and even short sentences in Braille. So far, this vision is flickering and crude, but eventual success would confirm that embryonic nerve cells can serve as a living bridge between the human brain and the computer.

Others, meanwhile, see the biochip going very much further than this. One biochip pioneer, Dr James McAlear of the US firm Gentronix, has suggested that a biochip could be a microprocessor constructed from organic molecules, like the human brain itself. He has argued that such molecular computers 'open up the possibility of three-dimensional circuits, increased speeds, reduced energy consumption, and ultraminiaturization that can reach a million billion elements per cubic centimetre. On this scale, all the memory elements of every computer manufactured to this day could be contained in a cube one centimetre on a side.'

Many scientists have tended to shrug off this vision of the future as the product of overheated imaginations in search of easy money, but Nature shows that it can be done. The leaf of a green plant, for example, contains 10 million more electronic elements per square millimetre than one of today's much-vaunted silicon chips.

McAlear and Professor Jacob Hanker of the University of Carolina took a small step in the right direction when they coated a glass slide with a layer of protein one molecule thick, later coating it with a thin protective 'resist' and etching a pattern between the resist elements with an electron beam. When the resulting chip was dipped into a silver solution, the protein organised the silver into microscopic circuitry which worked as well as a conventional microchip.

Ultimately, a molecular computer will have to be built up layer by layer. Instead of simply using the protein as a support for conventional circuitry, however, the elements of the computer might actually be grown from DNA templates genetically engineered into bacteria. The underlying concept of this type of biochip technology is sound enough. Pick, or construct, the right kind of molecule and it will adopt one of two distinct states when exposed to a minuscule electric charge. If you take one such state as 'zero' and the other as 'one', you have the basis of a binary code, the heart of most modern computer languages.

The original biochip work was carried out by two IBM scientists who synthesized the first biochip and received a patent in 1974. IBM, however, was not convinced and the project was shelved. But those who are driving the field of bio-electronics forward today believe that the ultimate limits of current chip technology are now in sight. Simply stated, the more you try and cram onto a chip, the thinner the 'walls' need to be between each of the electrical elements—and the more likely it is that electrons will burrow through those walls, with a consequent loss (or corruption) of information in the system.

At a time when the semiconductor field is seething with new ideas, such as magnetic bubble memories and Josephson's junctions, which exploit the phenomenon of superconductivity to cut resistance to electrical currents and thereby boost transmission speeds, it is easy to argue that silicon technology has such a start on the biochip that bio-electronics simply cannot catch up. But, as the human brain shows, organic molecules have one massive advantage over silicon chips: they can be organised into incredibly complex, three-dimensional arrays.

Just across the road from Gentronix stands the gleaming headquarters of a much larger genetic-engineering firm, Genex. Genex's vice-president for advanced technology, Dr Kevin Ulmer, has suggested that eventually molecular computers could assemble themselves, just as a growing organism does. 'One component could not assemble out of place or out of turn,' he argued, 'because it would lack the necessary binding sites required of the correct molecule. The yield of perfect devices could approach 100 per cent.' The final goal, he said, would be to 'develop a genetic code that will function like a virus, assembling a fully operational computer inside a cell'.

For the moment, this *is* science fiction, but computers are leading biotechnologists in some equally extraordinary directions which could become reality much sooner. Dr Sydney Brenner has modestly suggested that what genetic engineers do today is more like genetic mechanics than genetic design (see page 30), but a number of rapidly emerging technologies suggest that genetic design worthy of the name is only just over the horizon.

Screen Tests for Star Molecules

It may seem like an unlikely marriage of talents, but at least one major US drug company has been thinking of teaming up with film-makers such as George Lucas, producer of the *Star Wars* series, and Walt Disney Productions. Du Pont is not planning to fund the next Luke Skywalker adventure: instead, it has been wondering whether Disney or Lucas' company, Industrial Light and Magic, might be able to help in the design of new drugs.

The link is film animation. A growing number of drug companies are using computer-aided molecular-design techniques to get a better idea of what drug molecules look like, and of how they achieve their effect in the body. The goal is to work out how such molecules can be re-engineered to boost or modify their medical effects.

Sit down at one of the computer terminals used in this sort of work and you enter an astonishing world of swirling, swivelling patterns of multi-coloured light. In the latest 'vectorscope' displays, a minicomputer permits the user to call up 3D models of molecules. Once on the screen, structures containing up to 3,000 atoms can be rotated, so that the molecular biologist can get a clear idea of how a particular molecule hangs together. The images on the screen are formed 25 times a second, so that the human brain fuses them into a continuous, animated sequence. On the latest 'raster' terminals, which are more like a standard TV, the user can roam through the structures of molecules containing up to 15,000 atoms—and these models can show shading, colours (coding, say, for electrical charge), highlights and transparency. They can give the user an insider's view of even the most complex molecule and of the interplay between its various elements.

By using a computer joystick, you can rotate a particular structure or zoom in and out, as the moment demands. More than one structure may be displayed on the screen at the same time, so that they can be superimposed to check for striking similarities or differences. Spotting shared characteristics can help explain what makes a particular molecule taste sweet, or another, like endorphin, engage so precisely with a receptor in the human body.

A computer can display the shape of a molecule far more accurately than the old-style ball-and-stick models. Such accuracy can be extremely important, since small differences in shape can have a dramatic effect on a molecule's activity. As you rotate the molecule of the moment, keying in any structural changes you want to try making, the computer can calculate what the broader impacts on the structure might be.

Recently, scientists at another drug company, Merck, Sharp and Dohme, have been working on the hormone somatostatin, which may be useful in treating diabetes. By comparing different configurations of the hormone, all of them biologically active, they found that only 4 out of the 14 amino-acid groups in the molecule's structure were producing the desired effect. They then synthesised a very much simpler compound made up of just those four groups, and found that they had a hormone which had not only a stronger effect than the natural hormone but also a longer-lasting effect. Such products will have to be subjected to rigorous animal and human clinical trials lasting many years before they can be sold, but the potential is clear.

These new techniques have been spreading like wildfire through the drug industry. In the late 1970s, only four people at Du Pont were using computers in molecular research, but this figure had risen to about 250 by 1984 and was still growing rapidly.

The earliest pay-offs could well come in the enzyme business. One exploding area of interest is 'protein engineering', with Genex very much one of the leaders—despite the fact that Allied, the chemical company which had been sponsoring this work, pulled out of the joint venture the two companies had set up in 1983. In fact, brokers E. F. Hutton saw the terms of the severance agreement as highly advantageous for Genex, which 'will receive total cash of $12 million in one and one half years, and 100 per cent ownership of the fruits of the research projects, instead of $16.5 million over five years and only partial ownership of the R&D results'.

But what is Genex likely to be doing with this money? Useful clues can be found in the work that the UK's Laboratory of Molecular Biology has been doing. 'Ultimately, we should be able to alter an enzyme's surface to order,' predicted Dr Greg Winter following the announcement that he and his colleagues had succeeded in boosting one enzyme's catalysing ability (see page 29). Before you can begin to redesign a molecule, however, you have to have a detailed picture of its structure.

Until the advent of modern computer techniques, this was infernally difficult, often impossible. Indeed, it had taken ten years of painstaking work by another team of UK scientists to construct a three-dimensional model of this first enzyme to be remodelled. We need this sort of model because an enzyme can consist of anything up to 1,000 amino acids, typically found in 'side chains', arranged like ribs branching off from a central backbone. Try to change one of these side chains and you will almost certainly find that you have altered not only the enzyme's surface chemistry but also some of the very properties that made it commercially valuable in the first place.

Enzymes are marvels of biological complexity, but they are also often highly unstable. This means that even fine tinkering with their structure can deactivate or destroy them. This is critically important, since industrial enzymes have to be fairly robust to withstand the high-temperature/high-pressure environments in which many products are made.

The successful scientists, based at the Laboratory of Molecular Biology and at London's Imperial College, manipulated genes coding for two distinct amino acids in the enzyme's structure. They produced two novel enzymes as a result, one of which was twice as effective as the natural enzyme, the second 25 times as efficient. Unfortunately, this particular enzyme has no commercial value at present, but the Imperial College team has now switched to subtilisin, an enzyme used in biological detergents. 'Natural subtilisin only works at low temperatures,' said one of the Imperial College scientists. 'A version which was stable at higher temperatures would be commercially useful.'

The combination of computer-aided design techniques and protein engineering, which can involve changing the gene sequence of a molecule by means of a technique called 'site-directed mutagenesis', promises to be extremely powerful. And it has another attraction: you can patent a molecule which you have engineered into a novel configuration. Cetus and Genentech are among the genetic-engineering companies now moving rapidly into this field.

'A chemist can learn a lot more from a picture of a molecule than from an inch-high stack of numbers,' the head of Du Pont's computational chemistry department explained. But the most outlandish idea in this area probably came from Dr David Pensak, responsible for the company's molecular-modelling programme. He had been thinking of buying a $3-million pilot-training flight simulator for his research group. 'Conceptually, landing a 747 on an airfield strip is no different from landing a molecule on an enzyme-receptor site in the body,' he argued. 'We want our people to *feel* like drug molecules.'

Tempted to Play God?

Concern about the implications of genetic engineering has been voiced in many quarters during the last decade, beginning with the 1975 Asilomar conference on the hazards involved in recombinant-DNA technology. This conference followed the publication of the Berg letter, signed by many prominent scientists, which called for a voluntary moratorium on some classes of genetic-engineering work. However, the debate took a dramatic new turn in the summer of 1980 when President Carter received a letter penned on behalf of an unlikely alliance of Catholic, Jewish and Protestant church organisations. Genetic engineers who knew their ecclesiastical history, and recalled what had happened to pioneering scientists like Galileo and Darwin, read that letter very carefully indeed.

'We are moving rapidly into a new era of fundamental danger,' it began, 'triggered by genetic engineering. Albeit, there may be opportunity for doing good, the very term suggests danger.' Who, the letter demanded to know, should determine how human good is best served when new life-forms are being engineered? Who should control genetic experimentation, with all its untold implications for human survival? And who, it asked, would benefit—and who bear any adverse consequences, directly or indirectly?

'New life-forms may have dramatic potential for improving human life,' it admitted, 'whether by curing diseases, correcting genetic deficiencies or swallowing oil slicks. They may also, however, have unforeseen ramifications, and at times the cure may be worse than the original problem.' This line of thinking was developed several years later by such opponents of genetic engineering as citizen–advocate Jeremy Rifkin in the attempt to stop the release of genetically engineered bacteria designed to cut frost damage to crop plants (see page 199). Rifkin was by far the most effective of the biotechnology industry's critics, calling for new laws to control the release of new organisms. 'Industry didn't police itself during the petrochemical age,' he said, 'and it didn't police itself during the nuclear age. It's unrealistic to ask them to do it with biotechnology.'

Asked which other experiments he might challenge, Rifkin replied that any work which released genetically engineered micro-organisms into the environment would be suspect. 'The question,' he suggested to *Genetic Engineering News*,

> is, what is the difference between a living product and a chemical product in the environment? Here's the difference: a living product is inherently more unstable than a petrochemical product, merely by the fact that it's alive and interacts synergistically with other living things. Secondly, a living product can reproduce, migrate and grow. And third, a living product—micro-organism, bacteria, plant—cannot be recalled to the laboratory.

Agricultural microbes are not the only products of biotechnology which concern such critics, however. Others could be released from research laboratories, pollution-control plants and microbial leaching operations designed to extract metals from mine spoil. The recovery of copper from the drainage water of mines was almost certainly a widespread activity in the Mediterranean basin as early as 1000BC. It is known also that the leaching of copper on a large scale was well established in Spain by the 18th century. But what none of these miners realised was that they were being actively assisted by millions of microbial miners, which help convert copper into a soluble form that can be carried out of spoil heaps in the leach water.

Today, bacteria which naturally occur in spoil heaps, such as *Thiobacillus ferro-oxidans*, are deliberately exploited to recover millions of kilos of copper from billions of tonnes of low-grade ore. Copper obtained in this way accounts for more than 10 per cent of total US production. Similar techniques can be used to extract even a material such as uranium. Some companies, like the precious-metals producer Engelhard and photographic film-maker Kodak, are interested specifically in microbial silver recovery, whether from effluent from precious-metal refining or from waste X-ray film and other materials. Clearly, such micro-organisms are going to be improved by genetic engineers, where the economics of the exercise make sense—and, in some cases, even where they do not.

The sort of scenario which opponents of genetic engineering are fond of producing is exemplified by the following nightmare. The bacterium *E. coli*, which commonly inhabits the human intestine and is now the basic work-horse of genetic engineering, might be tailored to produce industrial alcohol. It could then escape, recolonise the human intestine, and create a world of unwitting inebriates.

The consensus among genetic engineers and most regulators, however, is that such fears have been enormously exaggerated.

Although many countries set up special bodies, like the UK's Genetic Manipulation Advisory Group, to police the industry, their work-load declined once the basic guidelines had been established. Indeed, when I visited Biogen's laboratories in Cambridge, Massachusetts, its ultra-secure P3 containment facility was being used as a temporary bicycle shed.

No, the issue that lurked at the heart of that letter to President Carter was deeper-rooted. Those who read it carefully knew that the escape of genetically engineered micro-organisms, already a widely debated issue, was not the central concern of the three church organisations. The real problems, the writers felt, would come when human beings began to modify human beings. 'History has shown us that there will always be those who believe it appropriate to "correct" our mental and social structures by genetic means, so as to fit their vision of humanity,' they warned. 'This becomes more dangerous when the basic tools to do so are finally at hand. Those who would play God will be tempted as never before.'

The Frankenstein Factor

President Carter's response to the letter was to set up a Commission for the Study of Ethical Problems in Medicine and Biomedical and Behavioural Research, which reported in 1983. Anyone who has ever tried to pin down the ethical or social implications of an emerging area of science and technology will sympathise with the Commission's conclusion, in *Splicing Life: The Social and Ethical Issues of Genetic Engineering with Human Beings*, that its main objective must be to 'stimulate thoughtful, long-term discussion rather than truncating such thinking with premature conclusions'. But it did recognise in its opening pages that public concern about gene splicing seems 'to reflect a deeper anxiety that work in this field might remake human beings, like Frankenstein's monster'. The media, of course, have been responsible for whipping up some of this concern. 'Simply put,' wrote Susan Carson in the *Winnipeg Tribune* in 1979,

you take a cell from some plant or animal and extract the chemical (DNA) that governs all the physical and mental characteristics of the whole being. Do the same with another, totally different, plant or animal. Graft the two together. Presto! Shake hands with an orange that quacks, with a flower that can eat you for breakfast—or even with the Flying Nun.

This may be 'simply put', but did it really help people understand what was going on?

A large part of the problem, however, stems from the facts that recombinant-DNA technology and cell-fusion techniques have been developed only fairly recently and are highly complex; and, as previous chapters have shown, their development is accelerating across a broad front of activity. 'What is remarkable about the science of gene splicing,' the Commission stressed, 'is not that it seems strange to laypeople—for all science is arcane to those who do not specialise in its study—but rather how unfamiliar it would be for the geneticists of even one generation ago.' Indeed, it added, perhaps 'the most predictable aspect of this technology may be its very unpredictability'.

Yet it is worth noting that human beings have intentionally set out to induce genetic changes in animals and plants for over 10,000 years, albeit often with little understanding of quite how they were achieving the desired effect. The practice of medicine, too, has already produced significant changes in our genetic make-up. The use of insulin to treat diabetes and the prescription of eye-glasses for myopia are examples of human ingenuity which have increased the prevalence in the population of genes which can have a deleterious effect in individuals.

As far as genetic engineering goes, the Commission distinguished between two broad medical applications: the use of drugs produced by means of genetic techniques, such as insulin or interferon, and the use of more direct approaches, including genetic counselling or gene surgery. The use of genetic screening and counselling has already enabled some groups of potential parents to make informed decisions about avoiding the occurrence of some genetic defects by terminating pregnancy, by artificial insemination or by avoiding conception altogether.

There was uproar when the lead-lined Repository for Germinal Choice, a supposedly bomb-proof sperm bank which accepts donations from Nobel laureates and top scientists, was established in the USA, but some of the gene-therapy and gene-surgery techniques which are now appearing on the horizon promise (or threaten, depending on your view) to be much more effective in remoulding the unborn.

The full-scale genetic engineering of people is so far in the future that we need hardly think about it, but genetic engineers are already working on techniques which will enable them to transplant genetically engineered bone-marrow cells into people suffering from such inherited diseases as Lesch-Nyham syndrome and PNP deficiency (which involves the shrivelling of the thymus gland, crippling the immune system, and leads to early death from infection), both of which stem from the lack of a single gene. The idea is that marrow cells would be extracted from the patient, the missing gene would be inserted and the modified cells would then be returned.

There are plenty of defective genes to 'edit' out of our genetic make-up. Our species suffers at least 3,000 genetic diseases, from blood disorders to the many forms of mental retardation. The signs of impending change are all about us. 'It's just a miracle,' said one Los Angeles woman who had been trying to become pregnant for over ten years. 'The miracle of this is not that I've given birth, but that someone else's egg has grown in my body.' New companies, like Fertility & Genetics Research, are springing up to meet demand in the booming 'baby business'. Inevitably, such companies are using the new genetic techniques to improve their service.

Most of us would agree that scrutinising the DNA involved in such genetic transfers makes sense, and new techniques are now being developed which will enable genes predisposing an unborn child to such diseases as Huntingdon's chorea and Down's syndrome to be detected much earlier in pregnancy— or perhaps even, ultimately, around the time when an egg is first fertilised in a test-tube. The use of gene probes and other techniques promises to identify any genetic predisposition to these diseases and to others which include breast or colon cancer, diabetes, depression, schizophrenia and a form of premature senility known as Alzheimer's disease. 'This new technology has an extraordinary power to predict any disease where there is any kind of genetic influence,' said Nancy Wexler, president of the Hereditary Disease Foundation. 'Instead of looking in a crystal ball to see your future, you'll look in your genes.' Well, yes and no. Even if such tests show up a genetic predisposition, say, to breast cancer, we are still dealing only with probabilities, not certainties. A daughter of a breast-cancer patient, for example, has a 10 per cent higher risk of developing breast cancer than other women. What sort of decisions would you take to avoid a 10 per cent higher risk?

The first attempt to treat a disease with human genes caused a massive furore. Dr Martin Cline, a professor of medicine at the University of California at Los Angeles, tried to correct a genetic blood disease by injecting normal genes into the bone marrow of two patients. They failed to function and the roof fell in on Cline, but similar experiments are certain to be attempted.

One possible problem, as parents begin to sidestep the 'natural lottery' of sexual reproduction by using *in vitro* fertilisation, surrogate mothers or frozen-embryo technology, is that our understanding of what a genetic 'defect' is will actually begin to slip. For example, if the genetic engineer can offer a means of guaranteeing that a child has a higher IQ, then today's 'normal' levels would certainly be considered deficient tomorrow. Our notion of what it means to be a 'good' parent would certainly change.

Where parents choose (or are forced) to take the gene-surgery option, the Commission suggested, their sense of family and kinship may also change radically. The use of gene surgery, involving operating on the fertilised egg itself, could obviously result in inheritable changes—and could encourage people to think of their family as extending much further into the future. Alternatively, they may simply conclude that the future development of such techniques will make a nonsense of traditional concepts of lineage anyway and proceed as non-aristocrats always have done.

Even relatively simple products of genetic engineering, such as human growth hormone (see page 82), could raise tricky ethical issues. No one doubts that growth-hormone therapy is desirable for many children who do not currently receive it. 'But,' asked Dr Thomas Murray, 'what if the child simply comes from short parents, and inherits shortness? What about a child who is not short at all, but whose parents believe (with some justification) that being taller still confers distinct social advantages?'

But even Rifkin admits that there is no way of regulating genetic engineering out of existence. Any country that tried to do so would simply force it underground—or overseas. The Commission's overall conclusion was that, while genetic engineering is not inherently inappropriate for human use, it is a painful irony that, in seeking to extend our control over our world, we often unwittingly lessen it. And, if there is one area where biotechnology could get out of hand in very short order, it is bioweaponry.

Beating Ploughshares into Swords

The idea, even soldiers admit, is ugly. Germ warfare, once described as 'public health in reverse', has a long, dishonourable history stretching back at least as far as the 14th century, when the Tartars catapulted plague victims into the besieged Crimean town of Kaffa. Far from beating its swords into ploughshares and its spears into pruning hooks, our species has displayed a unique facility for turning everyday tools into deadly weapons of war. 'The chlorine that poisoned our grandfathers at Ypres,' Robert Harris and Jeremy Paxman pointed out in *A Higher Form of Killing* (1982), 'came from the synthetic dye industry and was available thanks to our grandmothers' desire for brightly coloured dresses. Modern nerve gases were originally designed to help mankind by killing beetles and lice,' they continued, but in military hands they became 'insecticides for people'.

Now, with increasingly powerful biological tools emerging in the biotechnology industry, some genetic engineers are going public with their concerns about the military and terrorist potential of the new recombinant-DNA techniques. In itself, this is clearly a positive development, but it is chastening to recall that Albert Einstein similarly warned of the military potential of nuclear physics. If Einstein failed to deflect his colleagues from the path to Armageddon, can today's genetic engineers really hope to do any better?

Perhaps. Not because they are cleverer or more politically astute than Einstein, nor because peace is about to break out around the world, but because germ warfare is such an indiscriminate business. No one has yet developed a patriotic germ. On the other hand, some genetic engineers recognise that their new tools could help develop more reliable biological weapons, although few see any immediate prospect of 'ethnic weapons', able to exploit genetic differences between populations.

On April 4, 1972, the USA and the Soviet Union signed the Biological Weapons Convention, guaranteeing that they would 'never in any circumstances develop, produce, stockpile or otherwise acquire or retain' biological weapons. The Americans invited the press to watch as they incinerated containers of germs, or neutralised them with caustic soda. The Russians simply—incredibly—issued a statement that they had no bacteriological weapons. Worse, there was no provision in the Convention for independent, on-site verification of compliance—a vital requirement. While it is fairly simple to spot a nuclear plant from a satellite, a biological-warfare plant is a totally different proposition. The plant and equipment needed to produce chemical and biological weapons can also be used to produce more innocuous products, or may even be already producing the chemical or biological warfare agent but for non-military purposes. And many of these agents are astonishingly potent, so that small-scale batch production is all that is needed.

On the face of it, however, some of biology's swords were being beaten into ploughshares. Fort Detrick, the main US biological-weapons facility, was turned over to the National Cancer Institute. In the UK, the Public Health Laboratory Service took over the Ministry of Defence's Microbiological Research Establishment at Porton Down in 1979, renaming it the PHLS Centre for Applied Microbiology and Research. CAMR has since been building up its resources in healthcare biotechnology, focusing on the development of new and improved drugs, vaccines and diagnostic reagents.

Yet, even during the period of high optimism following the 1972 Convention, the UK and USA reserved sections of Porton Down and Fort Detrick to provide them with a 'watchtower capability'. The Convention does not ban continuing research for defensive purposes, they say. The trouble is that, if you want to develop and test a really effective defence against biological weapons, you may just find yourself producing the weapons you have agreed not to develop in order to test your preparedness. And, if your potential enemies get wind of that research, the biological arms race may begin again in earnest.

Paradoxically, in the run-up to the 1939–45 war, the Japanese took the fact that biological weapons had been banned in Europe as evidence that they must be effective. Both the Japanese and the Nazis tested such weapons on POWs and concentration-camp inmates, and the evidence suggests that the Japanese also used them against the Chinese, without much success.

By the end of the war, however, the USA's biological armoury was overwhelmingly superior and included anti-crop agents, which were seen as quicker-acting than a blockade and less contentious than atomic bombs. MIT biologist Dr Jonathan King recently pointed out that such crop destruction has been made immeasurably easier by the Green Revolution. Instead of triggering the modern equivalent of the Black Death, future biological warriors may opt for an updated version of the Irish potato famine.

The USA has accused the Soviet Union of using fungal weapons, in the 'yellow rain' controversy, while Cuba has accused the USA of introducing dengue and African swine fever into the country to ruin its economy. But there is no reason to suppose that the next wave of biological warriors will be Americans or Russians: the possibility that chemical and biological weapons were being used in the Iran–Iraq war in 1984 exercised experts as the conflict ground relentlessly, bloodily on.

One potential loophole in the 1972 Convention is the word 'lethal': the Americans spent a great deal of money researching psychochemicals, like LSD, behind a publicity screen of articles suggesting that it might be possible to 'eliminate death from war'. One government spokesman stated the objective like this: 'Ideally, we'd like something we could spray out of a small atomizer that would cause the enemy to come to our lines with his hands behind his back, whistling the Star-Spangled Banner.' In the event, however, the conclusion was that a Soviet general on LSD was just as likely to push the nuclear button.

The USA has published what it says is a complete list of the biotechnology research it is carrying out in the defence field. This ranges from vaccines against possible bioweapons, through biosensors to pick up the use of such weapons on the battlefield, to enzymes for use in decontaminating biological battlefields. Meanwhile, the US Defence Intelligence Agency was just one of the organisations claiming that the Soviet Union was ignoring the 1972 Convention, leading to calls for controls on biotechnology trade with the communist bloc, in the same way that computer and microchip exports were already controlled.

'The Soviet Union is assessed to have an active R&D programme to investigate and evaluate the utility of biological weapons,' reported an environmental biologist from the Agency in 1984. 'This effort violates the biological and toxin weapons convention of 1972, which was ratified by the USSR.' He noted also that there were 'at least seven biological warfare centres in the USSR that are under strict military control'.

The limitations of some biological weapons are illustrated by the continuing contamination of Gruinard Island, off the northwest coast of Scotland, after experiments made during World War II with bombs containing spores of *Bacillus anthracis*. CAMR has suggested that formaldehyde might be used as a possible decontaminant, but the task would be a formidable one, even now, since anthrax spores are virtually indestructible, hanging about in the soil until a suitable host wanders by.

Apart from anthrax, germ-warfare researchers have looked at the bacteria responsible for causing diseases such as brucellosis, cholera, glanders, plague, psittacosis, Rift Valley Fever, smallpox, tetanus, tularemia, typhoid, Venezuelan equine encephalitis and yellow fever. But, even if you vaccinate your front-line troops against such weapons, the chances are that some of the agents you are using will mutate in the wild, especially in battlefield conditions, and rebound on you. Genetic engineering could help make such weapons hit specific targets rapidly and then fade away into harmless by-products, but no one is claiming to have achieved this end yet.

An even more ominous possibility was suggested by the news that a Copenhagen teacher and his students had successfully transferred genes from one bacterium to another in a classroom experiment. The very messiness of biological weapons might give an unscrupulous group or government a surprising amount of political leverage.

And there is no need to think of the Libyas of this world: South Africa's National Institute for Virology resisted calls from the WHO that it should destroy its cultures of live smallpox virus. Only two laboratories are authorised to hold such cultures: the Center for Disease Control in Atlanta and the Institute of Virus Preparations in Moscow. With the elimination of smallpox in the wild and the falling immunity of the world's population to the disease, it is not difficult to imagine why South Africa might want to hold onto the cultures. The National Institute for Virology's director admitted that the decision 'was not taken for scientific reasons'.

Equally frightening is the possibility that terrorists or even a highly motivated individual might use recombinant-DNA methods to engineer a strain of bacteria or virus which was not only resistant to antibiotics or, let's say, interferon, but reacted to them by producing an even more powerful toxin. Science-fiction author Frank Herbert built an all-too-credible book, *The White Plague*, around this idea. Driven insane by an IRA outrage, a molecular biologist develops a lethal pathogen which seeks out and attaches itself to the human chromosomes in such a way that it kills only females. Incredible? Not as far as Herbert is concerned. 'After setting a basic cost for a relatively sophisticated lab at around $200,000 to $300,000,' he said later, 'I did a little more research. This was after the book was published. I found that, if you go into the surplus and used markets, my estimate was very high. I know of a surplus $6,000 centrifuge that went for $17.95.' Indeed, one of the reasons that BP failed to sell so much of the fermentation equipment from its failed Toprina plant (see page 176) was that it devoted a good deal of effort to making sure it did not get into the wrong hands. Others might not be so careful.

Genes for Industrial Growth

If you visit CAMR today and consult the visitor's book, you will get an excellent indication of the growing enthusiasm from around the world for biotechnology. 'There has been a lot of interest from foreign firms,' agreed CAMR director Dr Peter Sutton, 'with the Japanese particularly interested.' A few months later, CAMR became one component of a new industrial group, called Porton International, launched by UK businessman Wensley Haydon-Baillie.

With backing from 15 City institutions, including the pension funds of Barclay's Bank, Esso, ICI and the Imperial Group, the new group pulled together such UK companies as LH Engineering, renowned for its fermentation equipment, and Speywood Laboratories. Interestingly, however, Haydon-Baillie did not solicit venture capital because of his conviction that the new group is not a risk venture. Porton International, he explained, had been set up to exploit what he sees as the 'first commercial decade' of biotechnology.

Country after country, company after company, have announced plans to break into biotechnology. The developing countries, worried that they will be left out of the race, have lobbied for the International Centre for Genetic Engineering and Biotechnology, which they see as a means of plugging into the biotechnology revolution. 'Biotechnology, perhaps more than any other area of advanced science, offers solutions to the old problems which continue to hold many countries in the Dark Ages,' said Dr Burke Zimmerman of Cetus, who had served as a consultant to the UN Industrial Development Organisation during the planning phase for the new centre. These problems obviously include disease, malnutrition, overpopulation and the high cost and political vulnerability of energy imports.

It is still very difficult to predict what the long-term economic and employment impacts of biotechnology will be, however, although it is clear that there will be both gains and losses. But the potential for using modern techniques to improve existing Third World biotechnologies is clearly enormous, whether in agriculture, the preparation of fermented foods or healthcare.

Meanwhile, expect continuing controversy over the ownership and movement of the world's most valuable natural resource, the genes which contribute to the productivity and health of the major crop plants. In 1983, the UN Food and Agriculture Organisation was rocked by the fiercest debate in its history when Third World countries complained about the stranglehold which Western seed companies have gained on these plant-genetic resources, regardless of where they are found. Among the major companies involved are Cargill, Celanese, Ciba-Geigy, Occidental Petroleum, Sandoz and Shell. The fear is that such companies will engineer the crop plants they sell such that they do best when provided with inputs, such as fertilizers and pesticides, which only those companies can provide. There has been anger, too, at the activities of visiting botanists, with some Third World advocates protesting that 'we give them our plants, then they sell the seed back to us'.

And expect continuing controversy over the employment implications of biotechnology. Even such relatively developed countries as France and the UK often feel in the shadow of the USA, wondering whether they will be able to develop and maintain internationally competitive biotechnology industries. Biogen's recruitment of Dr Richard Flavell struck a particularly raw nerve in the UK because of concern about the biotechnology brain-drain to the USA. A study by the Institute of Manpower Studies found that the UK had lost 250 biotechnologists since the mid-1970s, and is continuing to lose highly qualified biotechnology specialists at the rate of about 30 a year—compared with a total of 1,500–2,000 biotechnologists working in the country's new biotechnology industries. But, due to the high level of automation in the industry, the Institute concluded, biotechnology is not going to provide many new jobs for university graduates in the near future, with perhaps an extra 100 a year needed into the 1990s.

Meanwhile, however, many companies which are already in the field, like Du Pont, are upgrading their existing biotechnology activities. When Du Pont opened its new $85-million biomedical and agricultural research facility late in 1984, executive vice-president Robert Forney stressed that 'our basic polymers businesses aren't dying. But when we look at relative growth rates, we feel we have to be big players in these new areas.'

The rate of growth in the employment of biotechnologists has obviously been much greater in the USA than in Europe or even in Japan, which many see as the main threat to the US dominance of the field. But will the big companies squeeze out many of the smaller start-ups as the field matures? Inevitably, there will be company failures, takeovers and mergers, but many of these smaller companies will find a niche. Often, they offer services which big companies would find it difficult to develop in-house.

'People like myself would never have gone and taken a job at a Becton Dickinson, or a SmithKline, or an Abbott,' explained Oncogene Sciences chief executive Dr John Stephenson. 'I don't want to be an employee of a large corporation.' In his own company, however, he is happy to work twelve hours a day, seven days a week. Becton, in fact, bought 20 per cent of the new company's equity when it was launched in 1983, reckoning that the development of its proposed range of cancer diagnostic kits could be advanced by a couple of years if it supported people who were already doing the work, as opposed to setting up its own operation from scratch.

A major strength of the new companies is that they are often addressing markets about which the bigger companies know very little, markets which require a combination of front-rank scientific skills and very fast commercial footwork. 'The markets that are easiest to get a grasp on are, of course, the markets that already exist,' as Biogen's Walter Gilbert put it. 'The products we see developing are new products which fill niches different than the ones already existing. We see ourselves as part of a new industry, not a replacement industry.'

For example, most people who think about agricultural bio-technology tend to think of new crop plants or such targets as biological nitrogen fixation (see page 119), but there are some very attractive targets much closer to hand. Take plant-disease diagnostics, which some biotechnologists suspect could be a $1-billion-a-year market by the early 1990s. 'We still talk of plant diseases in terms of stunts, rots, wilts, rusts and spots,' noted Agrigenetics president David Padwa. 'It's like we were still speaking of human medicine in terms of vapours, spirits, colours and humours.'

Previously available diagnostic techniques have often prod-uced results too late to be of much value to the farmer, largely because samples have to be taken to distant laboratories and run through time-consuming tests. Now new diagnostic kits are beginning to appear which are based on monoclonal antibodies, which can be used in the field and which give an immediate result. Some of the earliest targets have included long-lived crops, like grapevines and fruit trees, which represent an even greater investment than annual crops. And DNA Plant Tech-nology has had another clever idea: it plans to sell diagnostic kits to normal householders, for about $2 a time, so that they can test their lawns for half a dozen of the most common grass diseases.

So, while some biotechnologists may warn, like ICI director Dr Charles Reece has done, that 'we have taken the first few steps up a very long staircase', the overall picture, despite the darker side of the equation, is remarkably promising. 'There is no doubt that these new technologies, these new developments in the field of biotechnology, are going to have tremendous significance in the world,' as Dow Chemical's biotechnology director, Dr John Donalds, summed up the prospect. 'They are just beginning, and the fact that we don't have any industrial products today shouldn't bother any of us at all. We have an exciting field, with tens of thousands of new targets to pick and choose from. I am looking forward to the next few years.'

But if someone offers you shares in a new company being set up to develop and market 'Instant Passion' aphrodisiacs, ask for a copy of the prospectus—and subject it to very careful scrutiny indeed. You can still lose your shirt betting on biotechnology.

Further Reading

Many books and reports have been written on biotechnology, and many more will be written. I receive several new publications for review every week and know of a fair number in the pipeline. But some, inevitably, are better than others. If *The Gene Factory* has whetted your appetite for more information on biotechnology, you may find the following publications of interest.

Perhaps the shortest, although one of the most influential publications in Britain, was *Biotechnology: Report of a Joint Working Party*, better known after its Chairman, the late Dr Alfred Spinks, as the Spinks Report. This 63-page publication first appeared in 1980 and is available from Her Majesty's Stationery Office, London.

In the United States, two of the most important reports on biotechnology were produced by the Office of Technology Assessment. The first, published in 1981, was *Impact of Applied Genetics: Micro-Organisms, Plants, and Animals* (OTA-HR-132). The second, published in 1984, was *Commercial Biotechnology: An International Analysis* (OTA-BA-218). These were heavyweight reports, running to 331 and 612 pages respectively, and are available from the Office of Technology Assessment, Congress of the United States, Washington, D.C. 20510, USA.

The Organisation for Economic Co-operation and Development (OECD) also published a useful 84-page report, *Biotechnology: International Trends and Perspectives*, in 1982. This was written by Professors Alan Bull, Geoffrey Holt and Malcolm Lilly. The OECD is based in Paris, but some of its publications can be bought through Her Majesty's Stationery Office.

If you are looking for readable books on the science underlying biotechnology, there is considerable choice. Two books which should be on the shelves of any self-respecting student of biotechnology are *Man-Made Life: A Genetic Engineering Primer*, by Jeremy Cherfas (Basil Blackwell, Oxford 1982) and *Biotechnology: A New Industrial Revolution*, by Steve Prentis (Orbis, London 1984). Cherfas has been a regular contributor to *New Scientist*, while Prentis has been Editor of *Trends in Biotechnology* and *Trends in Biochemical Sciences*, both journals aimed at the specialist.

If, on the other hand, you want a highly readable book focusing on the social and political implications of biotechnology, try Edward

Yoxen's *The Gene Business: Who Should Control Biotechnology?* (Pan Books, London 1983).

The history of the debate about the safety of recombinant DNA technology has been described in a number of books. Two possible candidates are *The DNA Story* (W. H. Freeman, San Francisco and Oxford 1981), by J. Watson and J. Tooze, and *Genetic Alchemy: The Social History of the Recombinant DNA Controversy* (MIT Press, Cambridge, Massachusetts, and London 1982), by Sheldon Krimsky. Alternatively, if you want to view the debate exclusively from the opposition's side of the fence, read Jeremy Rifkin's *Algeny: A New World* (Pelican Books, London 1984).

Among the more business-oriented publications, one of the best was *The Business of Biotechnology*, by David Fishlock of the *Financial Times*— published by the FT in 1982. A shorter report on the implications of biotechnology for the British economy, *Biotechnology & British Industry*, was written by Peter Dunnill and Martin Rudd, and was published by the Swindon-based Science and Engineering Research Council's Biotechnology Directorate in 1984. More specifically, an excellent (if expensive) book on industrial enzymes is *Industrial Enzymology: The Application of Enzymes in Industry* (MacMillan/The Nature Press 1983), edited by Tony Godfrey and John Reichelt.

Most of these books contain bibliographies which will lead the reader deeper into the subject, but a number of the areas covered in *The Gene Factory* lie somewhat off the beaten track. If you are interested in microbial enhanced oil recovery, for example, see if you can find a copy of *Bacteria and the Enhancement of Oil Recovery* (Applied Science Publishers, London and New Jersey 1982), by V. Moses and D. G, Springham. If, on the other hand, you are intrigued by the implications of biotechnology for the developing countries, try *Microbial Processes: Promising Technologies for Developing Countries* (National Academy of Sciences, Washington, D.C. 1979) or *Blending of New and Traditional Technologies: Case Studies* (Tycooly International Publishing, Dublin 1984), which was produced by the International Labour Office.

If biotechnology's implications for the energy prospect are your main interest, read *The Alcohol Economy: Fuel Ethanol and the Brazilian Experience* (Frances Pinter, London 1983), by Harry Rothman, Rod Greenshields and Francisco Callé. A highly technical book on the microbial degradation of toxic chemicals, but one with extensive bibliographical notes, is *Biodegradation and Detoxification of Environmental Pollutants* (CRC Press, Baton Rouge, Florida 1982), edited by Ananda Chakrabarty.

A tremendously readable book on the future prospect in genetic screening and gene therapy is *Genetic Prophecy: Beyond the Double Helix* (Bantam Books, New York 1982), by Zsolt Harsanyi and Richard Hutton. A more technical account of some aspects of this area can be found in *The Role of Genetic Testing in the Prevention of Occupational Disease*, another report produced by the US Office of Technology Assessment (OTA-BA-194, 1983). An equally readable book on the darker side of biotechnology is *A Higher Form of Killing: The Secret Story of Gas and Germ Warfare* (Triad/ Paladin, London 1983), by Robert Harris and Jeremy Paxman.

For more regular briefing on developments in biotechnology, where books are less appropriate, a considerable number of periodicals and newsletters now cover various aspects of this burgeoning field. Rather than picking one information source to cover all eventualities, it may be helpful if I simply list some of the sources I find particularly useful when compiling *Biotechnology Bulletin* (published by OYEZ Scientific & Technical Services, London).

First, simply to illustrate the way in which rather more traditional publications have taken to reporting on biotechnology, I read four newspapers each day, each of which generates some material on biotechnology. These are the *Financial Times*, the *Guardian*, the *Times* and the *Wall Street Journal*. If I had to pick one of these four papers, however, it would probably be the *Financial Times*. The *Economist* is also an excellent source of information and comment, as are *Business Week* and *New Scientist*. For more specialised information, try *Biofutur* (Biofutur S.A., Paris, France), *Bio/Technology* and *Nature* (Nature Publishing Company, London and New York) and *Trends in Biotechnology* (Elsevier Science Publishers, Cambridge, UK).

There are also now a considerable number of biotechnology newsletters, the following of which I skim at the very least—and in some cases read almost cover to cover: *Agricell Report* (Agritech Consultants Inc, Shrub Oak, NY 10588, USA); *Applied Genetics News* (Business Communications Company, Stamford, CT 06906, USA); *Biotechnology News* (CTB International Publishing Company, Summit, NJ 07901, USA); *Current Biotechnology Abstracts* (Royal Society of Chemistry, Nottingham, UK); *Genetic Engineering and Biotechnology Monitor* (United Nations Industrial Development Organization, Vienna, Austria); *Genetic Engineering Letter* (Environews Inc, Washington, DC, USA); *Genetic Engineering News* (Mary Ann Liebert Inc., 157 East 86th Street, New York, NY 10028, USA); *Genetic Technology News* (Technical Insights Inc, Fort Lee, NJ 07024, USA); and *McGraw-Hill's Biotechnology Newswatch* (McGraw-Hill Inc, New York, NT 10020, USA).

Any reader still unable to find what he or she needs having scanned this list can try writing to me at the following address: John Elkington, Director, Bioresources Ltd, 10 Belgrave Square, London SW1X 8PH, England.